Remembering
EINSTEIN

Remembering EINSTEIN

Lectures on Physics and Astrophysics

edited by
B.V. SREEKANTAN

UNIVERSITY PRESS

YMCA Library Building, Jai Singh Road, New Delhi 110 001

Oxford University Press is a department of the University of Oxford.
It furthers the University's objective of excellence in research, scholarship,
and education by publishing worldwide in

Oxford New York

Auckland Cape Town Dar es Salaam Hong Kong Karachi
Kuala Lumpur Madrid Melbourne Mexico City Nairobi
New Delhi Shanghai Taipei Toronto

With offices in
Argentina Austria Brazil Chile Czech Republic France Greece Guatemala
Hungary Italy Japan Poland Portugal Singapore South Korea Switzerland
Thailand Turkey Ukraine Vietnam

Oxford is a registered trademark of Oxford University Press
in the UK and in certain other countries

Published in India by Oxford University Press, New Delhi

© Oxford University Press 2010

The moral rights of the author have been asserted
Database right Oxford University Press (maker)

First published 2010
Third impression 2011

All rights reserved. No part of this publication may be reproduced,
or transmitted in any form or by any means, electronic or mechanical,
including photocopying, recording or by any information storage and
retrieval system, without permission in writing from Oxford University Press.
Enquiries concerning reproduction outside the scope of the above should be
sent to the Rights Department, Oxford University Press, at the address above

You must not circulate this book in any other binding or cover
and you must impose this same condition on any acquirer

ISBN-13: 978-0-19-806449-7
ISBN-10: 0-19-806449-7

Typeset in Baskerville by Anvi Composers
Printed in India by De Unique, New Delhi 110 018
Published by Oxford University Press
YMCA Library Building, Jai Singh Road, New Delhi 110 001

Contents

Foreword — vii
Acknowledgements — ix
Introduction by B.V. Sreekantan — xi

1. VIRENDRA SINGH — 1
   Albert Einstein: His *Annus Mirabilis* 1905

2. ARVIND KUMAR — 22
   Einstein and Light Quanta

3. JAYANT VISHNU NARLIKAR — 35
   Einstein and Cosmology

4. SHASHIKUMAR MADHUSUDAN CHITRE — 48
   Role of Relativity in Astronomy and Astrophysics

5. THANU PADMANABHAN — 60
   Cosmology and Dark Energy

6. SANDIP P. TRIVEDI — 73
   Einstein's Dream and String Theory

7. ABHAY VASANT ASHTEKAR — 85
   Space and Time: From Antiquity to Einstein and Beyond

8. NARESH DADHICH — 97
   Why Einstein (Had I been born in 1844!)?

9. B.N. JAGATAP — 111
   Bose-Einstein Condensation: When Atoms become Waves

List of Contributors — 146

Foreword

The year 2005 had been designated the 'International Year of Physics' by UNESCO to mark the 100th anniversary of Albert Einstein's *Annus Mirabilis*. In 1905 Einstein wrote five remarkable papers which transformed the whole landscape of science by ushering the quantum revolution in physics.

With a view to capture the exciting developments that have occurred during the course of the past century in the wake of Einstein's landmark contributions, the Nehru Centre, Mumbai organized a commemorative lecture series by inviting distinguished scientists to highlight Einstein's work and its far-reaching impact on contemporary developments in physics. The lectures covered a wide range of problems dealing with atoms and light quanta, special and general relativity and their role in astrophysics and cosmology, Bose-Einstein condensation, string theory, and dark energy.

We hope that this collection of essays will provide a glimpse of the excitement at the frontiers of modern science that Einstein's phenomenal contributions initiated over a century ago.

SHARAD PAWAR
Chairman
Nehru Centre, Mumbai

Acknowledgements

This is a compilation of the lectures delivered at the Nehru Centre by eminent astrophysicists in 2005 to commemorate the centenary of the papers published by Einstein in 1905. Hence the title *Remembering Einstein*. The year 2005 was also the 'International Year of Physics'. The lecture series was conceptualized by Shri Piyush Pandey and Shri Suhas Naik-Satam of the Nehru Planetarium. We appreciate their initiative in organizing the lectures.

I am grateful to Shri Sharad Pawar, Chairman, Nehru Centre for writing the Foreword to this volume. My special thanks are also due to Professor B.V. Sreekantan who has edited this book and written an Introduction to it. Our librarian, Mrs Arati Desai has shouldered the entire responsibility of getting this volume published. Many thanks to her.

I.M. KADRI
General Secretary
Nehru Centre

Introduction

In 1905 Albert Einstein wrote five papers, and all of them were of such fundamental nature that together with Max Planck's quantum hypothesis and a spate of experimental discoveries in the last decades of the nineteenth century, they completely transformed the course of modern physics, astrophysics, and the philosophy of science. Naturally, the year 2005 was celebrated as the year of physics commemorating the centenary of the *Annus Mirabilis*—'the year of miracles'—all over the world, especially by the physics community. The celebrations were of many types. The Nehru Centre at Mumbai organized a series of lectures on the works of Einstein and their impact on and implications for physics and astrophysics. The lectures were delivered by eminent scientists mostly from Mumbai and Pune. This book contains the nine lectures delivered at the Nehru Centre for a popular audience, mostly students.

The inaugural lecture, 'Albert Einstein: His *Annus Mirabilis* 1905' was delivered by Virendra Singh, former Director of the Tata Institute of Fundamental Research, on 9 July 2005. Singh started with a reference to the Annus Mirabilis of Issac Newton—the year 1666, and the contributions of Newton in the two plague-hit years in England, 1665 and 1666—calculus, dynamics, optics, gravitation, and so on. He was still an undergraduate in Cambridge at that time.

After making a reference to the early life of Einstein as a clerk in the Patent office, Singh offered a brief description of the five papers written during 1905—the first on 'Production and transformation of light', the second on 'Molecular sizes' (Ph.D. dissertation), the third on 'Brownian motion', the fourth on 'Special theory of relativity, and the last one on '$E = mc^2$'. Though these belonged to widely different areas of physics, they had a tremendous impact on the course of modern physics in the twentieth century. The presentations by Singh are incisive, elegant, and technically accurate, and bring out Einstein's unique role in the unfolding of relativity and quantum mechanics. Singh points out that the extent of greatness of Einstein was aptly reflected in the famous cartoon of Herblock in *The Washington Post* in which planet earth was identified by the words 'Albert Einstein lived here'. According to *Time* magazine survey, Einstein was the 'Man of the Millennium'.

The second lecture was by Arvind Kumar, Director of the Homi Bhabha Centre for Science Education, Mumbai on 'Einstein and Light Quanta' in which he elaborates on the route Einstein took to resolve the problems of black body radiation that had baffled the scientific community of the late nineteenth and early twentieth centuries, and to which Max Planck had made seminal contribution by introducing the quantum hypothesis though with considerable reluctance. The hesitation and scepticism about the quantum hypothesis was shared by many senior scientists of the time. But Einstein, after explaining the photoelectric effect for which he got the Nobel Prize, felt that there was no doubt about the reality of light quanta. This was also later confirmed by the observation of the Compton Effect. Proceeding further, Kumar discusses the emergence of Bose-Einstein and Fermi-Dirac statistics. Though light quanta have become part of the received wisdom in physics for the past hundred years, it is strange that Einstein was uncomfortable with the notion of light quanta till the end of his life. In spite of all his work, he felt he still did not understand what the 'photon' was.

The third lecture in the series was by Jayant V. Narlikar, former Director of Inter-University Centre for Astronomy and Astrophysics (IUCAA), Pune. In his lecture, 'Einstein and Cosmology', Narlikar starts with a historical account of the knowledge of the universe emphasizing how even in early twentieth century when Einstein began his work on general relativity and its bearing on cosmology, the centuries-old idea of a 'static universe' was very much there and influenced Einstein in framing his general theory of relativity. The first version of Einstein's equations provided solutions which could not be fitted into the notion of a static universe, and precisely for overcoming this problem he introduced a parameter 'λ'. Later, he regretted introducing this parameter when observationally it became clear through the work of Hubble that the universe was not static, but was expanding. Narlikar discusses the pros and cons of big bang and steady state theories, the discovery of microwave radiation, the physics of the early hot universe leading to the new field of astroparticle physics. Referring to the recent topics of dark matter and dark energy, he cautions against the speculative nature of these and emphasizes the need for alternative theories and more definitive observations.

The fourth lecture was by S.M. Chitre, retired Senior Professor of Tata Institute of Fundamental Research (TIFR), on 'Role of Relativity in Astronomy and Astrophysics'. Chitre's lecture focuses on the properties of the collapsed condensed objects in the sky, the neutron stars and black holes which have been recognized as sources of very high energy phenomena discovered after the advent of radio astronomy and space astronomies. Of particular interest are quasars, pulsars, x-ray stars, x-ray binaries, gamma ray bursters, and so on. He discusses the formation of these remnants of the supernovae explosions of very massive stars. Illustrating the power of modern technologies in observational astronomy, he points out that in the old days the observed 43 seconds per century precession of the planet Mercury was being taken as evidence for predictions of the general theory of relativity. The recent observation of the progressive twisting of the elliptical orbit of the compact binary 1913 + 16 by as much as 4.2 degrees per year reveals this same gravitational effect much more convincingly. Further, the change in the pulsar period of one part in 10^6 per year is evidence of the emission of gravitational radiation by this system. Chitre provides a fascinating account of the interplay of observation and theory that is currently enriching the field of physics and astrophysics, influenced by the intricacies of the special and general theories of relativity of Einstein.

We have seen earlier that in the last decades of the nineteenth century and the early decades of the twentieth century, experimental discoveries and theoretical formulations led to a total transformation of our basic concepts of space, time, mass, energy, and even of the universe. Classical physics had to give place to relativity and quantum mechanics and static universe to one of expanding universe. While the series of discoveries and theories made during the twentieth century with technological advances enabled observational investigations of higher sensitivity and resolution, the last decade of the twentieth century came up with another major discovery: the accelerating universe, a discovery which has rocked the physics and astrophysics communities and the prospects of yet another new physics emerging is not ruled out. T. Padmanabhan, Senior Professor at IUCAA, Pune highlights this new development in his lecture. He begins with a presentation of the current ideas on the cosmos which though consisting of billions of galaxies can be looked upon as smooth and homogeneous and isotropic on scales of 200 Megaparsecs and larger. Though the universe as a whole is expanding, everything in it on the micro-scale is not expanding. Close to the big bang which started the universe off, it was infinitely small, smaller than an electron even. The very high temperature ($10^{28°}$K) cooled to about $3000°$K after 40,000 years, and now after 13.6 billion years, the temperature is $3°$K.

Recent observations seem to lead to the theory that nearly 70 per cent of the energy is in the form of dark energy; 26 per cent in the form of dark matter; and only four per cent in the form of matter and radiation we are familiar with. How do we understand this composition? What is dark matter? What is dark energy? How does dark energy bring about acceleration of the expanding universe? Is it the same as the 'λ' that Einstein had introduced to make the universe static? These are the challenging issues which Padmanabhan discusses.

In 'Einstein's Dream and String Theory', Sandip Trivedi of the Tata Institute of Fundamental Research presents an enthusiastic but cautious perspective on the current status of the string theory and how the framework of this theory may fulfil Einstein's dream of a unified theory which implied the formulation of a single set of laws on the basis of which all phenomena in nature can be explained. Einstein spent thirty years of his later life searching for this unified theory but did not succeed. The problem was one of fitting gravitation into quantum mechanics. Also, it became clear from the developments in high-energy physics and the standard model that the ultimate constituents of all matter are particles. With the advent of the string theory in the 1980s, the idea of particles was replaced by strings in the sense that even particles—electrons, quarks, photons, etc.—are made of 'strings' and the different particles correspond to the different modes of vibration of the same kind of string. It is important to note that this is only a hypothesis which has no experimental support yet. A further assumption that is made is that the strings interact in the framework of quantum mechanics and this leads to gravitation and other forces. The string theory requires extra dimensions. Next, Trivedi comes to the question of the accelerating universe and the hypothesis of dark energy characterized by a repulsive gravitational force and its explanation in terms of the cosmological constant λ originally introduced by Einstein. Trivedi points out that his own work along with Shamit Kachra, Ranata Kallosh, and Andre Linde and based on the earlier work of Prasanta Tripathy has shown that string theory can lead to a positive cosmological constant λ resulting in the acceleration of the universe. However, the main disappointment is that the string theory is not able to give the observed rate of acceleration of the universe. A wide range of values is possible. These are the challenges for the future.

In 'Space and Time: From Antiquity to Einstein and Beyond', Abhay Ashtekar of the Institute of Gravitational Physics and Geometry, Penn State University, and an associate of IUCAA, Pune, and Institute for Theoretical Physics, Utrecht, starts with a historical account of the concepts of space and time which provide a stage on which the drama of all interactions unfold. Aristotle's paradigm of absolute time, absolute space, and absolute rest frame identified with earth prevailed till 1686, when Newton toppled this centuries-old dogma. The privileged status of the earth as the absolute rest frame was removed in accordance with the Galilean principle of relativity, retaining the absolute notion of time. Strangely, Clarke Maxwell's electromagnetic theory, while specifying an absolute value for the speed of light (c), was without an absolute frame of reference. An absolute speed 'c' contradicted the Galilean relativity principle. The ether was thought to be the reference frame for Maxwell's equations by most physicists of the time. It is in this context of a perplexing situation that Einstein came out with his special theory of relativity in which only four-dimensional space-time continuum had an absolute meaning. Time intervals and spatial intervals by themselves depended on the choice of a reference frame. Ashtekar points out that this new paradigm resulted in dramatic predictions like the equivalence of mass and energy ($E = mc^2$). Also in the middle of the nineteenth century itself, mathematicians had realized that space may not be Euclidean and may be curved due to the presence of matter. Galileo had demonstrated through his leaning tower of Pisa experiment that gravity is universal: all bodies fall the same way under the action of gravity. Moreover, one cannot build gravity shields. These ideas led Einstein to formulate his general theory of relativity and to state that gravity is not a force but a manifestation of space-time geometry. Planets like the earth move in this curved geometry of space.

The magic of general relativity, Ashtekar emphasizes, is that through an elegant mathematics it transforms these conceptually simple ideas into concrete equations that enable astonishing predictions to be made about physical reality. Over the last three decades many of the physical consequences predicted have been experimentally verified.

Edwin Hubble's observation of the expanding universe fitted very well into the solutions of Einstein's equations by the Russian mathematician Alexander Friedmann, and the big bang theory of creation led to the idea of the universe having a beginning when the density of matter and curvature of space-time were infinite. The discovery of microwave radiation in the 1960s confirmed this scenario which meant space, time, and matter all had a beginning. A relevant question is whether the big bang is a frontier where all physics stops?

The question of stellar collapse and formation of black holes is discussed by Ashtekar in its historical perspective. Though the black hole was a result of the solution to Einstein's equations by Schwarzschild, Einstein himself had hesitation in accepting black holes. In modern astronomy, black holes are playing a prominent role in the explanation of many exotic phenomena.

At the big bang and black hole singularities, the world of very large and of the very small meet. Ashtekar points out that these singularities are our gates to reach beyond general relativity and once again revise dramatically our notions of space and time. In general relativity, space-time is modelled by a continuum. This breaks down at the Planck interval of 10^{-33} cm. One must use quantum space-time, one version of which is loop quantum gravity. Quantum geometry provides the mathematical tools and physical concepts to describe quantum space-time. Ashtekar then describes his own pioneering work in this field.

He talks about a forward-in-time motion picture of the universe in which there is a contracting pre-big-bang branch well described by general relativity. When the matter density is $0.8\ \rho_{planck}$ the repulsive force of quantum geometry becomes dominant, the contraction stops and gives place to a bounce, and joins on to the post-big-bang expanding branch. The pre- and post-big-bangs are joined by a quantum bridge governed by quantum geometry. These new ideas have been helpful in the attempts to resolve some of the longstanding problems in cosmology and black hole physics.

In 'Why Einstein (Had I been born in 1844!)?', Naresh Dadhich, Director of IUCAA, Pune, provides a fascinating account of how using just principles of universalization and recognizing the universal character of universal entities, Einstein's theory of relativity follows and how even Maxwell's theory of electromagnetic waves with a constant velocity of light automatically follows. As an extension of the same principles he ponders over profound questions: How many dimensions does gravity live? How many basic forces are there in nature? What are the building blocks of space-time?

One of the crucial questions is: Should vacuum be completely inert physically, or should it have some physical properties? Not very different from the questions one asked about ether in the old days. He points out that observationally what is established is (i) electromagnetic waves propagate through vacuum and (ii) we cannot measure any motion relative to vacuum. It cannot act as a *reference frame*. To satisfy these observations vacuum has to have a microstructure (like matter) which are the building blocks of space. Quantum fluctuations are the basis for all happenings. What is it that fluctuates? Space has to acquire quantum behaviour at the micro level—this is what universality requires. But what are the building blocks of space? Nobody knows. Both quantum theory and general relativity require a new quantum theory of space-time that is valid in quantum-space-time.

B.N. Jagatap of the Laser and Plasma Division of the Bhabha Atomic Research Centre, spoke about one of the most fascinating topics associated with two great scientists of the twentieth century, Albert Einstein and Satyendra Nath Bose: 'The Bose-Einstein Condensation'. The formation of Bose-Einstein condensates was predicted in 1925, but was experimentally realized only in 1995 since it required cooling of atoms to temperatures as low as a few hundred nano Kelvin—a technological challenge. Jagatap has presented a vivid account of the intricacies of the various methods that were developed in this connection: evaporative cooling, laser cooling, etc.

The wave-like properties of atoms become dramatically apparent through the Bose-Einstein condensation phenomenon, and the entire assembly of atoms behaves like one giant matter wave with all the atoms oscillating in phase. This coherent matter wave behaves entirely like a laser. All the associated phenomena in lasers like interference, diffraction, and amplification become apparent. In 'atomic optics', as it is known, the matter waves are manipulated by electric and magnetic fields. Jagatap also discusses new developments in the areas of atom lasers, non-linear atomic optics, atom amplifications and also atomic interference, quantum computers with cold atoms, etc. This is yet another supreme example of how a purely theoretical prediction of a possible new state of matter was realized experimentally after sustained effort and opened up new possibilities of technological applications.

It is hoped that these nine lectures, elaborating the work of Albert Einstein in the early twentieth century and projecting the tremendous influence that Einstein's work has had on many fundamental issues in physics and astrophysics and also their fallouts in technology, will enrich the minds of readers, young and old, and particularly those who could not attend the lectures delivered at the Nehru Centre. The Nehru Centre richly deserves our congratulations for arranging these lectures and also for bringing them out in the form of a book. Authored by scientists who are actively engaged in research in these very fields, this will certainly be a valuable addition to a topic which, though a century old, continues to have powerful influence on current physics and astrophysics.

1

VIRENDRA SINGH

Albert Einstein[1]
His *Annus Mirabilis* 1905

YEAR OF PHYSICS (2005)

The year 2005 was celebrated by UNESCO as 'International Year of Physics'. It was the centenary year of the Annus Mirabilis or the Miracle year 1905 of Albert Einstein. During this year he published a set of five papers dealing with the existence of atoms, special relativity including the now famous equation $E = mc^2$ expressing the equivalence of energy content E and inertial mass m of a body, as well as on the quantum theory together with its application to the photoelectric effect.[2] These papers mark the watershed between classical physics of Isaac Newton, Michael Faraday, and John Clerk Maxwell and modern physics. It is therefore entirely appropriate that this centenary year of the Annus Mirabilis was celebrated as the year of physics.

We may also mention that the decade of 1895 to 1905 was extremely rich in discoveries which established the existence of a number of phenomena which were not explicable within classical physics. This crisis would require for its resolution a change of classical framework to that of modern physics involving relativity and quantum theory. In 1895, Roentgen discovered X-rays. In 1896 radioactivity was discovered by Henri Becquerel and magnetic field effect on spectral lines by Pietr Zeeman. In 1897, J.J. Thompson established the existence of electrons. In 1900, Max Planck introduced quantum ideas in physics. As has been noted in this connection, the twentieth century in physics began not in the year 1900, but a full five years earlier in 1895. It is also fitting that this fruitful decade (1895–1905) was capped by the Annus Mirabilis of Einstein.

Before we proceed to discuss in detail why the year 1905 is referred to as the Annus Mirabilis, let us briefly recall an earlier year, 1666 which is also referred to by the same designation. It was the Annus Mirabilis of Isaac

Einstein in 1905

During this year, Einstein published five papers on statistical physics, special theory of relativity, and the quantum theory apart from completing his Ph.D. dissertation. In chronological order these were as follows:

1. Light quantum paper: The paper 'On a heuristic point of view concerning the production and transformation of light' was received by *Annalen der Physik* on 18 March 1905. This was published in *Annalen der Physik*, 17, 132–48, 1905.
2. Thesis on molecular sizes: The Ph.D. dissertation 'On a new determination of the molecular dimensions' was completed on 30 April 1905. It was printed at Bern and submitted to University of Zürich on 20 July 1905. He also sent a paper based on the thesis to *Annalen der Physik* soon after the thesis was accepted on 19 August 1905 by the University which appeared in *Annalen der Physik*, 19, 289–305, 1906.
3. 'Brownian motion' paper: The paper 'On the motion of small particles suspended in liquids at rest required the molecular kinetic theory of heat' was received on 11 May 1905 for publication and appeared in *Annalen der Physik*, 17, 549–60, 1905.
4. Special theory of relativity paper: The paper 'On the electrodynamics of moving bodies' was received for publication on 30 June 1905 and appeared in *Annalen der Physik*, 17, 891–921, 1905.
5. $E = mc^2$ paper: The paper 'Does the inertia of a body depend on its energy content?' was received for publication on 27 September 1905 and appeared as *Annalen der Physik*, 18, 639–41, 1905.

Besides the above, Einstein sent another paper on 'Brownian motion' on 19 December 1905 to *Annalen der Physik* which was published next year. We had mentioned earlier Newton's own account of his Anni Mirabilis some half a century after the event. We have an account by Einstein of his work in his Annus Mirabilis which was written during that very year. It occurs in his letters which he wrote to his friend Conrad Habicht. He, together with Einstein and Maurice Solovine, was a member of the triumvirate 'Olympia Academy', who used to meet regularly in evenings to have wide ranging intellectual discussions extending from philosophy to physics.

Einstein wrote to Habicht on 18 or 25 May 1905:

> ... But why have you still not sent me your dissertation? ... I promise you four papers in return, the first of which I might send you soon, since I will soon get complimentary reprints. The paper deals with radiation and the energy properties of light and is very revolutionary as you will see if you send me your work first. The second paper is a determination of the true sizes of atoms from the diffusion and the viscosity of dilute solutions of neutral substances. The third proves that, on the assumption of molecular theory of heat, bodies of the order of magnitude 1/1000 mm, suspended in the liquids, must already perform an observable random motion that is produced by the thermal

motion; in fact physiologists have observed (unexplained) motions of suspended small, inanimate, bodies, whose motion they designate as 'Brownian Molecular motion'. The fourth paper is only a rough draft at this point, and is an electrodynamics of moving bodies which employs a modification of the theory of space and time; the purely kinematical part of this paper will surely interest you . . ..

Einstein again wrote to him between 30 June 1905–22 September 1905 to bring him up-to-date with his later work as follows:

A consequence of the study on electrodynamics did cross my mind, namely, the relativity principle, in association with Maxwell's fundamental equations, requires that the mass be a direct measure of the energy contained in a body; light carries mass with it. A noticeable reduction of mass would have to take place in the case of radium. The consideration is amusing and seductive; but for all I know, God Almighty might be laughing at the whole matter and might have been leading me around by the nose.

We have excised the purely personal remarks and banter from these letters.

In the rest of the write-up we shall now discuss these contributions in somewhat more detail and provide their background and context so as to appreciate them more properly. We shall not follow the chronological order in which they were written, but rather the order in which they make a transition from classical physics to modern physics. Chronologically his light quantum paper is first during 1905, but as Einstein himself remarked it is the most revolutionary. The order would therefore be as follows:

1. Thesis on molecular motion
2. 'Brownian motion' papers
3. Special theory of relativity and $E = mc^2$ papers
4. Light quantum papers

THESIS ON MOLECULAR SIZES

The First Attempt

The PhD thesis which Einstein wrote in 1905 was not his first attempt at submitting a thesis for this degree. He first submitted a PhD dissertation in November 1901, but the topic was not known. It was also not clear why Einstein withdrew it in February 1902. As he wrote to his friend Michele Besso from Bern on 22 (?) January 1903: 'I have recently decided to join the ranks of Privatdozenten, assuming, of course that I can carry through with it. On the other hand, I will not go for a doctorate, because it would of little help to me, and the whole comedy has become boring.' He, however, changed his mind about a doctorate degree soon afterwards.

the laboratory of Jean Perrin, Einstein requested his student and collaborator, Ludwig Hopf, to check his calculations again. Hopf was successful in finally finding the missing factor of $5/2$ in Einstein's expression. Hopf's correction was communicated to Perrin by Einstein on 12 January 1911. If this correction is used then we get the much more satisfactory value

$$N_A = 6.56 \times 10^{23}$$

Initially this dissertation was foreshadowed by other papers of Einstein during this year. However, this is the paper of Einstein which has received the highest citation in view of its use by molecular physicists and chemists. Maybe the citation index is not such an infallible guide to the significance of a paper!

BROWNIAN MOTION

Atomic Theory Around the End of Nineteenth Century

Modern chemistry dates back to John Dalton's book *New System of Chemical Philosophy* in 1808, in which he proposed his system of a finite number of chemical elements. All the molecules were taken as composed of atoms of these chemical elements. Amedeo Avogadro in 1811 proposed that, under conditions of equal temperature and pressure, equal volumes of gases contain the same number of molecules for all gases. This number for a mole of gas was named by Jean Perrin as the Avogadro Number N_A. Avogadro's Law presupposes the reality of molecules. Most chemists, however, used atomic theory in the nineteenth century as a theoretical heuristic device to bring order into the description of chemical phenomenon. They did not necessarily subscribe to their reality.

In the second half of the nineteenth century, the kinetic theory of gases, which posited the gases to consist of moving molecules, made rapid progress. Clausius, in 1857, suggested that heat is a form of molecular motion. John Clerk Maxwell proposed his famous Distribution Law for the molecular velocities in a gas.

Ludwig Boltzmann gave his equation which set out to reduce all thermodynamic phenomena to mechanical description using molecules. These developments in the kinetic theory of gases gave a big boost to the atoms being real entities. At the end of the nineteenth century, most physicists and chemists thus either believed in the reality of molecules or at least were willing to use them as heuristics.

In view of the fact that all the evidence for the atoms was indirect, as atoms were not directly seen, there was still a small but powerful opposition to the idea of their reality. The great physical chemist Ostwald, as well as George Helm, regarded atoms to be mathematical constructs. The situation in regard to atoms was similar to that of 'quarks' as constituents of matter in the twentieth century. Ostwald had his own programme, 'Energetics', in which the prime ontological entity was energy.

Max Planck was also of that persuasion at that time, since he regarded laws of thermodynamics to be absolute laws. While in Boltzmann's atomic view the second law of thermodynamics, regarding entropy, was only statistical in nature and not absolute. Even the great physicist and philosopher Ernst Mach was opposed to the reality of the atoms in view of 'positivist' slant of his philosophy. In 1905 Einstein made a decisive impact on this debate through his paper on 'Brownian motion'.

Einstein's Contribution[4]

In his paper 'Brownian Motion', Einstein investigated the random motions executed by visible, but very small, particles in a liquid. The visible random motion of these particles was taken to arise from their being buffeted by the incessant motion of the invisible liquid molecules. His hope was that such a study would be convincing enough about the reality of the underlying molecules of the liquid. He noted,

It will be shown in this paper that according to molecular Kinetic Theory of heat, bodies of a microscopically visible size suspended in liquids must, as a result of thermal molecular motion, perform motions of such magnitude that they can easily be detected by a microscope.

He continues, 'It is possible that the motions to be discussed here are identical with the so-called "Brownian molecular motion"; however, the data available to me on the latter are so imprecise that I could not form a definite opinion on this matter.'

Robert Brown, the English botanist, had observed random motion of pollen grains in a liquid in 1828. The motion was analogous to a drunkard's walk around a lamp post.

Brown as a result of his experiments ruled out the possibility that the observed motion was due to pollen grains being moved by some vital force, that is, due to their living nature. Many different suggestions such as effect of capillarity, role of convection currents, evaporation, and interaction with light and electrical forces were put forward to explain these random motions. Even kinetic theory was proposed as a possible explanation but Von Nageli, in 1879, ruled it out for reasons which appeared cogent. He took straight segments on the path of a Brownian particle to be their free motion between two collisions with molecules. We now know that even these straight segments arise due to the effect of multiple collisions with atoms. In fact, one of the primary achievements of Einstein in this paper was to clarify the physically significant observations to make on these particles.

Einstein calculated the diffusion constant D for the suspended microscopic particles, of the size a of the order of one-thousandth of a millimetre, in the liquid and showed that it is given by

$$D = RT/(6\pi \eta a N_A)$$

Two Clouds on the Horizon

Lord Kelvin, in a very perceptive and insightful lecture before the Royal Institution in April 1900 talked about two 'Nineteenth century clouds over the dynamical theory of heat and light'. One of these involved the continued unsuccessful attempts to experimentally measure the motion of the earth through 'luminiferous aether'. The other one of these referred to the failure of equipartition of energy in classical statistical mechanics.

The rest of Einstein's work during the miraculous year 1905 is devoted to dispelling these two ominous clouds hovering over the horizon of classical physics. His papers on special theory of relativity deal with a resolution of 'earth's velocity through aether' puzzle. This involves a complete overhaul of classical concepts of space and time. His paper on the 'light quantum' deals with the other cloud and ushered in the quantum revolution. We now turn to these papers.

SPECIAL THEORY OF RELATIVITY

Galilean Relativity

Newton's laws of motion are valid in a set of special frames of reference. These are called 'inertial frames of reference'. For example, Newton's first law says that a mass point, not acted upon by any external force, keeps moving in a straight line with a uniform speed. Now a particle which is moving in such a fashion in an earth-laboratory will not appear to move in a straight line when viewed from the sun due to earth's daily rotation and its annual revolution around the sun. Clearly the two frames of reference, that is, one in which the earth is at rest and other one in which the sun is at rest cannot both be inertial frames of reference.

How do we know if some particular frame of reference is inertial? We first note that if a frame of reference S is inertial then any other frame of reference S' which is moving in a straight line with uniform velocity is also inertial. This specifies the class of frames of reference which are inertial and in which Newton's three laws of motion hold. In order to characterize the class of the inertial frame we have to specify at least one of them. In practice, for solar system applications, it was taken to be the frame in which the centre of mass of the solar system is at rest or in uniform rectilinear motion. Within the accuracy required in these calculations, it was the same as the one in which the centre of mass of the universe was at rest or uniform linear motion or the one in which the system of fixed stars was at rest or uniform linear motion.

The rules for comparing the space and time coordinate measurement in different inertial frames are known as Galilean transformations. Newton's laws obey the principles of Galilean relativity. Their form is invariant, that is, unchanged, under Galilean transformation between two inertial reference systems.

Maxwell's Electromagnetic Theory and Galilean Relativity

Note that as long as Newton's laws of motion are the only fundamental laws of physics, there is no way in which one can determine the absolute velocity of any inertial frame with respect to some absolutely fixed point at rest. This situation radically changes with the advent of Maxwell's equations for electromagnetism.

We note that Maxwell's equations do not have the same form in different inertial frames connected by Galilean transformations. That is, they are not invariant under them. For example, the velocity of electromagnetic waves (say, light) is a constant. One can ask, in which inertial frame is it so? It can not be so in all inertial frames which are connected by Galilean relation. If it is given by $\vec{c}$ in its direction of propagation S, it would be $\vec{c} + \vec{v}$ in the frame S' which is moving with a velocity $\vec{v}$ rectilinearly with respect to S. The velocity of light was thus a fixed constant c only in the frame in which the luminiferous aether is at rest.

This clash between invariance of Newton's laws and noninvariance of Maxwell's electromagnetic theory of light opens a way by which the motion of earth, for example, can be experimentally measured with respect to universal aether. A large number of methods were thought for this purpose. All of them gave a null result. Experiments were unable to detect the motion of the earth through aether. The most celebrated and accurate experiment devised for this purpose was by Michelson and Morley in 1887, which also reported a null result. As Maxwell summarized in an article in *Encyclopaedia Britannica*, 'The whole question of the state of the luminiferous medium near the earth, and of its connection with gross matter, is very far as yet from being settled by experiment.'

Einstein's Resolution: Special Theory of Relativity

Einstein's resolution of 'earth-aether velocity' problem was obtained by a thorough revision of Newtonian concepts of absolute space and absolute time. In this revision he was guided by his analysis of the concept of simultaneity. If the two events take place in a single frame of reference, for example, a railway platform or a uniformly moving railway train on linear tracks, there is no difficulty in saying whether the two events are simultaneous in the same single frame of reference. If you start thinking about the problem one finds that the two events which look simultaneous in one frame, say railway platform, are not so in another relatively moving frame—that of a moving train. This is because the light signals used to observe the two events, whose simultaneity we are discussing, will take different times in the frames of two relatively moving observers. This is due to the speed of light signal being finite. Since simultaneity is not an invariant concept it follows that time cannot be absolute.

Einstein wished to hold on to what is now known as the two postulates of his special relativity theory. These are:

which completely absorbs all the radiation falling on it. It was also shown that the radiation inside a heated cavity is same as black-body radiation. Max Planck occupied Kirchhoff's chair at Berlin in 1889. He argued that as the ratio is independent of the nature of the cavity material he should be able to calculate it by using a simple model for the material of the cavity. The model he used was that it is made of Hertzian oscillators each with a single frequency ν. Using this model he could show that the universal function is related to average energy of each oscillator of frequency n, at the temperature T, of the black-body radiation. He had this result on 18 May 1899.

If Planck had known the equipartition theorem of classical statistical mechanics, for average energy, at this point, he would have obtained the law of black-body radiation now known as Rayleigh-Jeans Radiation Law as it was given by Rayleigh in June 1900 and corrected for a missing factor of 8 by Jeans in June 1905. Indeed this was first done by Einstein in his light quantum paper. Amusingly, he did it before Jeans. Rayleigh-Jeans Radiation Law was found applicable only at small values of ν/T and not for large values of ν/T. One thus became aware of the second cloud on the horizon of classical physics referred to by Lord Kelvin, namely, the failure of equipartition of energy.

Guided by the precision experimental results on black-body radiation Planck announced an empirical Radiation Law on 19 October 1900 which fitted the data perfectly. Planck's Radiation Law is now known to be the correct law of black body radiation. It had the same form as Rayleigh's Law for small ν/T and the form of empirically proposed Wien's Law, given in 1894, for large ν/T. Planck, however, had no theoretical basis for his Radiation Law.

Planck next presented a derivation of his Radiation Law. He was so desperate that he even used Boltzmann's probability interpretation for entropy. The derivation was announced to the German Physical Society on 14 December 1900. The new element in his derivation was his assumption that a Hertzian oscillator, of frequency ν, can emit or absorb radiation only in integral multiples of a basic quantum of energy ε, where $\varepsilon = h\nu$. The constant h is now known as Planck's constant. In classical physics there was no such discreteness. The oscillator could emit or absorb radiation of any energy. This was the first parting of ways with classical physics. Planck, however, took this assumption as a purely formal one and did not quite realize that something radical has been introduced. As he said 'This was a purely formal assumption and I really did not give it much thought except that no matter what the cost, I must bring out a positive result.'

Einstein's Light Quantum Hypothesis[8]

Einstein was the first person to realize that Planck's introduction of energy quanta was a revolutionary step. As we noted, Einstein in his light quantum paper first showed that the 'Rayleigh-Jean's Law' is the unambiguous prediction of classical physics for the Radiation Law. This law not only does not work for high frequency radiation, it also theoretically suffers from 'ultraviolet catastrophe' (that is, infinite energy). This convinced Einstein that to get the correct Radiation Law, a break with classical physics is involved.

In his quest for the cause of the failure of classical physics Einstein was guided by his unhappiness with asymmetrical treatment of matter and radiation in classical physics. Matter is discrete and particulate while radiation is continuous and wave-like. He thus proposes that radiation is also particle-like just as matter is (his 'light quantum' hypothesis) and not wave-like. He was of course fully aware of the successes of wave theory in dealing with the phenomena of interference and diffraction of light. All these phenomena, however, need only time averages and for such phenomena wave theory probably is indispensable. It is, however, conceivable that a basic particle picture on time averaging could produce wave like behaviour. He summarized that the particle nature of radiation may show up in the processes involving the generation and transformation of light where we deal with instantaneous processes.

Can one adduce any evidence in favour of the particle nature of light? Einstein proceeds to show that a consideration of Wien's Radiation Law, valid in the non-classical regime of large frequencies, does that. He calculates the probability p that the monochromatic radiation of frequency ν, occupying a volume V_0, could be confined to smaller volume V using Wien's Law. The result is

$$p = (V/V_0)^n \text{ with } n = E/h\nu$$

where E is the total energy of the radiation. This is of the same form as for a gas of n particles. From this remarkable similarity, Einstein concludes 'Monochromatic radiation of low density (within the range of validity of Wien's radiation formula) behaves thermodynamically as if it consisted of mutually independent energy quanta of magnitude $R\beta\nu/N$.' (In modern notation $R\beta\nu/N$ reads as $h\nu$). This is Einstein's light quantum hypothesis. In this picture

the energy of light is discontinuously distributed in space ... when a light ray spreading from a point is not distributed continuously over ever increasing spaces, but consists of a finite number of energy quanta that are localised in points in space, move without dividing, and can be absorbed or generated only as a whole.

Einstein applied successfully his light quantum hypothesis to other phenomena involving generation and transformation of light. The most important of these was his treatment of photoelectric effect. He also discussed Stokes' rule in photoluminescence and ionization of gases by ultraviolet light.

The Photoelectric Effect

In 1887, Heinrich Hertz observed that the ultraviolet light incident on metals can cause electric sparks. In 1989, J.J. Thomson established that the sparks are due to emission of the electrons. Phillip Lenard showed in 1902 that this phenomenon, now called the photoelectric effect, showed 'not the slightest dependence on the light intensity' even when it was varied a thousand-fold. He also made a qualitative observation that photoelectron energies increased with the increasing light frequency. The observations of Lenard were hard to explain on the basis of electromagnetic wave theory of light. The wave theory would predict an increase in photoelectron energy with increasing incident light intensity and no effect due to increase of frequency of incident light.

In Einstein's light quantum picture, a light quantum, with energy $h\nu$, on colliding with an electron in the metal, gives its entire energy to it. An electron from the interior of a metal has to do some work, W, to escape from the interior to the surface. We, therefore, get the Einstein photoelectric equation, for the energy of the electron E,

$$E = h\nu - W$$

Of course, the electron may lose some energy to other atoms before escaping to the surface, so this expression gives only the maximum of photo-electron energy which would be observed. One can see that Einstein's light quantum picture explains quite naturally the intensity independence of photoelectron energies and gives a precise quantitative prediction for its dependence on incident light frequency. It also predicts that no photoelectrons would be observed if $\nu < \nu_0$ where $h\nu_0 = W$. The effect of increasing light intensity should be an increase in the number of emitted electrons and not on their energy. Abram Pais has called this equation as the second coming of Planck's constant.

Robert A. Millikan spent ten years testing the Einstein equation and he did the most exacting experiments. He summarized his conclusions as well as his personal dislike of light quantum concept, 'Einstein's photoelectric equation ... appears in every case to predict exactly the observed results... yet, the semi-corpuscular theory by which Einstein arrived at his equations seems at present wholly untenable' (1915) and 'the bold, not to say reckless hypothesis of electromagnetic light corpuscle' (1916).

Envoi

Einstein was awarded the Nobel Prize in Physics (1921) for this paper on light quanta and especially its application to the photoelectric effect. Even though his status as public icon is associated with his relativity theory, he was not awarded Nobel Prize. He, however, delivered his Nobel lecture on Relativity.

Einstein's light quantum was renamed as 'photon' by G.N. Lewis in 1926. Though Einstein talked about photon energy, $E = h\nu$, in 1905, it is curious that he introduced the concept of photon momentum, $p = h\nu/c$, only in 1917. As we have seen, even Millikan did not believe in photons around 1915–16 despite his detailed experimental work on photoelectric effect. In 1923, the kinematics of the Compton Effect was worked out on the basis of its being an elastic electron-photon scattering by A.H. Compton successfully. After that it was widely accepted that light does sometimes behave as photon.

Einstein made the first application of quantum ideas to matter in his work on specific heat of solids in 1907. A consideration of energy fluctuations, using Planck's Radiation Law, led him to the dual particle-wave nature of radiation in 1909. In 1916–17 in the course of a new derivation of Planck's Radiation Law, using chemical kinetics methods, Einstein discovered the phenomenon of stimulated emission of light and introduced his famous A and B coefficients. These are of fundamental importance in the theory of lasers.

In 1924, S.N. Bose sent Einstein a new derivation of Planck's Law in which only the photon concept was used, albeit the photons did not obey the classical statistics of Maxwell and Boltzmann, but rather a new statistics. Einstein saw the importance of this contribution, translated the paper into German, and got it published. He also applied it to the material particles. The new statistics is now known as either Bose-Statistics or as Bose-Einstein Statistics. The particles which obey this statistics are known as bosons. As a consequence of these statistical considerations Einstein discovered that a free gas of bosons undergoes a phase transition, Bose-Einstein condensation, below a critical temperature. The phenomenon was seen only in 1995 and a Nobel Prize awarded for it in 2001.

The modern mathematical formulation of quantum mechanics was obtained by W. Heisenberg and E. Schrödinger in 1925-6 in two different versions, namely, matrix mechanics and wave mechanics. They both led to the same physical predictions and were thus physically equivalent. Einstein's role in achieving this transition from the old quantum theory to modern quantum mechanics was quite significant. He has been called the godfather of Schrödinger's the wave mechanics and his relativity theory with its emphasis on operational procedure provided the inspiration to Heisenberg in his matrix mechanics.

After 1926 Einstein's focus shifted to foundational questions of quantum mechanics. He gave his ensemble interpretation of quantum mechanics. His discovery of nonlocal correlations in quantum mechanics with Podolsky and Rosen in 1935 was of far-reaching significance and continues to spawn new fields, such as quantum computing, quantum information theory, and quantum cryptography, down to the present time.

CONCLUSION

Einstein purchased in 1935 a white simple farm house at 112, Mercer Street within walking distance of his office at Institute for Advanced Studies, Princeton and he lived here till the end. There were a few pictures and etchings. These included a drawing of Gandhi, whom he admired greatly. There were photographs of his mother and his sister Maja who lived with him after she moved to Princeton from Italy in 1939. He had also brought with him, from Europe, three etchings of physicists he admired more than any other. These were Newton, Maxwell, and Faraday. It is now abundantly clear, given his contributions to physics, that he himself belonged to this select company.

When Einstein died the famous cartoonist Herblock published in *The Washington Post*, a cartoon, in which the planet earth is identified by the words 'Albert Einstein lived here'. He became a world icon in 1919 and since then he has continued to hold a place of high esteem in public mind for his science, his humanity, his fight for a peaceful world, and his freedom from cant. At the end of the millennium he was voted by *Time* magazine survey, and many others as the 'Man of the Millennium'.

NOTES

1. The literature on the science and life of Einstein is enormous. The best biography for the physicist is A. Pais, *Subtle is the Lord: The Science and Life of Albert Einstein*, Clarendon Press, Oxford and Oxford University Press, New York, 1982. A volume for a more general reader is J. Bernstein, *Einstein*, Viking Press, New York, 1973.

2. For the writings of Einstein, we have the multivolume ongoing series, *The Collected Papers of Albert Einstein*, Princeton University Press, Princeton, NJ, 1987 and the companion volumes *The Collected Papers of Albert Einstein: English Translation*, Princeton University Press, Princeton, NJ, 1987.

    The brief quotations from Einstein's papers and letters and thesis reports used in this lecture are from Vol. 2: *The Swiss Years: Writings, 1900–1909*, Vol. 5: *The Swiss Years: Correspondence, 1902–1914*. Einstein's papers from the miracle year 1905 are also available in English translation in J. Stachel (ed.), *Einstein's Miraculous Year: Five Papers that changed the Face of Physics*, Princeton, 1998. (Indian reprint by Srishti Publishers, New Delhi, 2001.)

3. The quote from Isaac Newton is from R.S. Westfall, *Never at rest, A Biography of Isaac Newton*, Cambridge, 1981.

4. On Einstein's doctoral thesis, see also N. Straumann: arXiv: physics/0504201 (April 2005).

5. For an appreciation of Einstein's contribution to the theory of random processes, see L. Cohen, 'The History of Noise', *IEEE Signal Processing Magazine*, 20–45, November 2005.

6. Ostwald's quote on the reality of atoms is from Bernstein's book (p. 186) referred earlier.

7. For the impact of Einstein's work on twentieth-century Physics, see *Physics World*, Special Einstein Issue, January 2005 and *Current Science*, Special Einstein Issue, 25 December 2005.

8. For more detailed write-up on Einstein's contributions to quantum theory, see Singh (2005). [There is some overlap of the present write-up with this paper.]

REFERENCES

Beck, A. (1987). *The Collected Papers of Albert Einstein: English Translation*. Princeton, NJ: Princeton University Press.

Bernstein, J. (1973). *Einstein*. New York: Viking Press.

Cohen, L. (2005). 'The History of Noise', *IEEE Signal Processing Magazine*, 20–45 (Nov).

Giulini, D. and N. Straumann (2006). *Stud. Hist. Phil. Mod. Phys.*, 37, 115–73.

Pais, A. (1982). *Subtle is the Lord: The Science and Life of Albert Einstein.* Oxford: Clarendon Press and New York: Oxford University Press.

Singh, V. (2005). 'Einstein and the Quantum', *Current Science*, B9, 2101–12.

Stachel, J. (ed.) (1987). *The Collected Papers of Albert Einstein.* Princeton, NJ: Princeton University Press.

Straumann, N. (2005). 'On Einstein's Doctoral Thesis', see also: arXiv: physics/0504201 (April).

Westfall, R.S. (1981). *Never at Rest, A Biography of Isaac Newton.* Cambridge: Cambridge University Press.

2

ARVIND KUMAR

Einstein and Light Quanta

PRELUDES TO PLANCK'S LAW

Max Planck is usually regarded as the father of quantum mechanics, but depending on your persuasion, you could regard the birth date of quantum mechanics to be 19 October 1900, when Planck announced the black-body radiation law; or 14 December 1900, when he gave its theoretical derivation. Now, if there is a father, there needs to be a grandfather too! The best candidate for the latter is Gustav Kirchhoff who in 1859 proved a wonderfully simple result based on the then new science of thermodynamics. The same year Charles Darwin's famous book *The Origin of Species* was published. While the theory of evolution by natural selection soon triggered a revolution in biology, Kirchhoff's result was a precursor to a revolution in fundamental physics that unfolded only four decades later.

Consider a simple situation of thermal equilibrium: radiation in a cavity with its walls maintained at a certain temperature. This radiation is isotropic, homogeneous, and unpolarized. Further, energy density of the cavity radiation per unit frequency interval is independent of the size and shape of the cavity or the material of the cavity walls. These remarkable facts can be proved by arguing that if they were not true, the Second Law of Thermodynamics would be violated.

Let $\rho(\nu, T)\Delta\nu$ be the energy density of radiation in an isothermal enclosure (at temperature T) in the frequency interval ν and $\nu + \Delta\nu$. Kirchhoff's result can be summarized by saying that $\rho(\nu, T)$ is a universal function of ν and T.

There is nothing more appealing to physicists than universality. Remember Einstein's words that he was not interested in small details, in the spectrum of this or that element; he was interested in how 'HE thinks'. Well, universality is thought to be one vital key to unravel how 'HE thinks'. If $\rho(\nu, T)$ is universal, the challenge was to find its form empirically and justify it theoretically. In a sense, the challenge was truly met only with the discovery of quantum statistics by Satyendra Nath Bose in 1924.

But on the way there were many a milestone. First, Stefan in 1879 conjectured on the basis of experimental data that the total radiative power emitted by a body is proportional to the fourth power of its absolute temperature T. In 1884, Ludwig Boltzmann gave a thermodynamic proof of this law that is true only for black bodies. The proof appears in standard texts. But, remember, besides the Second Law, it makes use of the result of Maxwell's electromagnetic theory that radiation pressure is one-third the energy density. Stefan-Boltzman Law, as it is now called, may be written in terms of energy density U thus

$$U(T) = \int_o^\infty \rho(\nu,T)d\nu = aT^4 \tag{1}$$

The constant of proportionality is related to the well-known constant (σ), called the Stefan-Boltzmann constant, by the relation $a = \dfrac{4\sigma}{c}$, where c is the speed of light.

Thermodynamic reasoning can sometimes yield surprising results. In 1893, Wien constructed an argument using adiabatic expansion of cavity radiation, and arrived at the following scaling law:

$$\rho(v,T) = v^3 f(\dfrac{v}{T}) \tag{2}$$

Here f is a universal function of only one argument ν/T. In 1896, he conjectured a form for the function f, suggested by Maxwell's velocity distribution for an ideal gas.

$$\rho(v,T) = \alpha v^3 e^{-\beta v/T} \tag{3}$$

All along, experimentalists were making progress in refining their techniques to measure radiation and extending their frequency domain. Experiments by Paschen (1897) and by Rubens and Kurlbaum (1900) showed that Wien's Law, which worked well in the high frequency limit, seemed to be failing for lower frequencies. The experimental progress in the far infrared regime was crucial to this discovery.

In early 1900, there appeared another candidate for black-body radiation law. Using classical equipartition theorem for each mode of radiation, Rayleigh arrived at the law

$$\rho(\nu,T) = \dfrac{8\pi\nu^2}{c^3}kT \tag{4}$$

The correct proportionality factor above was in fact due to Jeans and came much later (1905); eq. (4) is, therefore, known as Rayleigh-Jeans Law. $\rho(\nu,T)$ diverges as ν becomes large, so Rayleigh put in an extra cut-off factor. But the law worked only in the low frequency limit—just where Wien's Law failed, and vice versa.

Planck knew of the failure of Wien's Law in the low frequency region. Perhaps stimulated by an afternoon tea conversation with Rubens and

Kurlbaum on a fateful day in October 1900 (in which he came to know that $\rho\alpha T$ for low ν (if not the full law, eq. [4]), he brilliantly interpolated between the two limiting forms of black-body radiation law and announced the following law on 19 October 1900:

$$\rho(\nu, T) = \frac{8\pi\nu^2}{c^3} \frac{h\nu}{e^{h\nu/kT} - 1} \qquad (5)$$

where h is now known to be a fundamental constant of nature called the Planck's constant. It has the dimension of energy × time or action or angular momentum. Clearly, for $\frac{h\nu}{kT} \gg 1$, $\rho(\nu, T)$ reduces to Wien's Law (eq. [3]); for $\frac{h\nu}{kT} \ll 1$, $\rho(\nu, T)$ reduces to Rayleigh-Jeans Law.

PLANCK'S THEORETICAL 'DERIVATION' OF BLACK-BODY RADIATION LAW

Planck's law (eq. 5) agreed with experimental data very well. Indeed the values of h and k obtained from the experimental fit of eq. (5) then are surprisingly close to their modern accepted values. Planck had no choice but to 'derive' eq. (5). He did what many of us do when faced with a difficult problem in an examination—he proceeded to work backward!

(i) First, he related $\rho(\nu, T)$ to the average energy $U(\nu, T)$ of the charged material oscillator of the wall. A charged oscillator of natural frequency ν (damped due to radiative energy loss) and forced by an external frequency ν' satisfies the equation

$$m\ddot{x} + m(2\pi\nu)^2 x - \frac{2e^2}{3c^3}\dddot{x} = eF\cos 2\pi\nu' t \qquad (6)$$

For small damping, $\dddot{x}$ may be replaced by $-(2\pi\nu)^2 \dot{x}$ and eq. (6) becomes the usual textbook equation for a forced damped harmonic oscillator.

The steady state energy E of the oscillator is then

$$E = \frac{e^2 F^2}{2m} \frac{1}{4\pi(\nu - \nu')^2 + \gamma^2} \qquad (7)$$

where γ is the damping constant given by $\gamma = \frac{8\pi^2 e^2 \nu^3}{3mc^3}$

If the external radiation is thermal radiation at temperature T, all that is needed is to replace F^2 by $\rho(\nu', T) d\nu'$ apart from some constant factors, and integrate eq. (7) over ν'. For small γ this yields

$$\rho(\nu, T) = \frac{8\pi\nu^2}{c^3} U(\nu, T) \qquad (8)$$

where $U(\nu, T)$ is the equilibrium energy of the oscillator at temperature T. Eq. (8) is known as Planck's link, connecting the energy density of radiation to the average energy of the oscillator in equilibrium. Using this link and

Planck's Law for $\rho(\nu, T)$ we get

$$U(\nu, T) = h\nu/(e^{h\nu/kT} - 1) \tag{9}$$

For low frequencies, $U(\nu, T) \to kT$, but failure of this classical equipartition theorem result is evident at high frequencies. Putting $U(\nu, T) = kT$ in eq. (8) yields the Rayleigh-Jeans Law (eq. [4]):

$$\rho(\nu, T) = \frac{8\pi\nu^2}{c^3} kT$$

Planck either did not notice all this or did not care. It was left to Einstein to realize this right in the beginning of his light quantum paper, and further to exploit eq. (9) in his 1906 work on specific heats. But Planck had a different agenda. He wanted to derive his formula for ρ; equivalently, *using* Planck's link, this amounted to deriving eq. (9).

(ii) But Planck had no way to derive eq. (9) directly. So he went yet another step backward. Inverting eq. (9), we get

$$\frac{1}{T} = \frac{k}{h\nu} \ln\left(\frac{U + h\nu}{U}\right) \tag{10}$$

Using the thermodynamic law $TdS = dU$, with T given by eq. (10) and integrating for a given ν,

$$S = k\left[\left(1 + \frac{U}{h\nu}\right) \ln\left(1 + \frac{U}{h\nu}\right) - \frac{U}{h\nu} \ln \frac{U}{h\nu}\right] \tag{11}$$

(iii) Thus Planck had converted the black-body radiation problem into a problem of the entropy of material oscillators. He had to somehow get eq. (11), for then he could retrace the steps and derive his experimentally successful formula for $\rho(\nu, T)$. He found himself doing an 'act of desperation':

Consider a large number N of oscillators, with total energy $U_N = NU$ and entropy $S_N = NS$

Now Planck made the 'wild' postulate that the total energy is made of finite energy elements ε.

$$U_N = P\varepsilon \tag{12}$$

Further, *a la* Boltzmann, he wrote

$$S_N = k \ln W_N \tag{13}$$

where he identified W_N as the number of ways in which P indistinguishable energy elements can be distributed over N distinguishable oscillators. This gives

$$W_N = \frac{(N - 1 + P)!}{P!(N - 1)!} \tag{14}$$

Putting this in eq. (13) and making use of the Stirling approximation, we get

$$S = k\left[\left(1 + \frac{U}{\varepsilon}\right) \ln\left(1 + \frac{U}{\varepsilon}\right) - \frac{U}{\varepsilon} \ln \frac{U}{\varepsilon}\right] \tag{15}$$

which reduces to the desired expression (11) for S if
$$\varepsilon = h\nu \qquad (16)$$

Planck's method of counting W_N was only superficially similar to Boltzmann's counting procedure—it did not then have a sound basis. Yet in conjunction with the quantum postulate, eq. (16), it worked and a new paradigm emerged in physics.

EINSTEIN'S ROUTE TO THE LIGHT QUANTUM HYPOTHESIS

In March 1905, Einstein completed the paper 'On a heuristic point of view concerning the generation and conversion of light' which contained the light quantum hypothesis and its application to, among other things, the photoelectric effect. The same year—Einstein's miraculous year whose centenary we are celebrating—he completed four more papers, besides his Ph.D. thesis. These latter papers included his discovery of special relativity and his greatly cited work on Brownian motion. Each of these three strands of his work in 1905 perhaps projects out a different profile of Einstein's creative personality. In the March paper that he himself regarded as revolutionary, Einstein was at his heuristic best.

The March paper opens with the realization that there seemed something fundamentally wrong with classical physics, for the basic results of its two sub-disciplines (classical mechanics and classical statistical mechanics) in the context of thermal radiation combined to give an absurd result, namely that the Stefan-Boltzmann constant is infinite. We saw this in our earlier discussion. Planck's classical link (eq. 8) and the equipartition result $U = kT$ yield the Rayleigh-Jeans Law (eq. [4])

$$\rho(\nu, T) = \frac{8\pi\nu^2}{c^3} kT$$

which disagrees with experiment for high ν and when integrated over ν gives a divergent integral. Einstein was confronted with a peculiar situation: a theoretically sound law (R-J Law) without experimental support and an experimentally correct law (Planck's Law) without a sound theoretical basis. (Planck's 'derivation' did not seem convincing.) This is where he hit upon a heuristic.

Wien's Law, as seen earlier, is valid at high frequency—where the classical Rayleigh-Jeans Law fails. So, Einstein guessed, Wien's Law is likely to contain some essential non-classical features. Since Wien's Law is much simpler to handle, why not examine the thermodynamics of Wien's radiation? Forget the more correct Planck's Law for a start and do not use the Planck's link. But what thermodynamic feature would he look for in his analysis of Wien's radiation? Here was another guess. In his earlier work a year before on energy fluctuation in radiation, he had found the volume dependence of entropy a significant thing to deal with. He did the same here. He studied the volume dependence of entropy of Wien's radiation and found it analogous to the volume dependence of entropy of a classical ideal gas of material particles. The analogy led him to the light quantum hypothesis, as follows:

Let $\phi(\nu, T)$ be the entropy per unit volume per unit frequency interval. The familiar relation $TdS = dU$ in terms of ϕ and ρ reads,

$$T^{-1} = \frac{\partial \phi}{\partial \rho} \tag{17}$$

For ρ, use Wien's Law

$$\rho = \alpha \nu^3 e^{-\beta \nu / T} \tag{18}$$

Inverting eq. (18), we get

$$T^{-1} = \frac{1}{\beta \nu} \ln \frac{\alpha \nu^3}{\rho} \tag{19}$$

Substituting eq. (19) in eq. (17) and integrating (with fixed ν) yields

$$\phi = -\frac{\rho}{\beta \nu} \ln(\frac{\rho}{\alpha \nu^3} - 1) \tag{20}$$

Now the entropy S and energy E between ν and $\nu + d\nu$ are

$$S(\nu, V, T) = \phi V d\nu \tag{21}$$

$$E(\nu, V, T) = \rho V d\nu \tag{22}$$

To study volume dependence, keep ν and E fixed

$$S(\nu, V, E) - s(\nu, V_o, E) = \frac{E}{\beta \nu} \ln \left(\frac{V}{V_o} \right) = k \ln \left(\frac{V}{V_o} \right)^{E/h\nu} \tag{23}$$

Compare this with the result for a classical ideal gas of n material particles:

$$S(V, E) - S(V_o, E) = k \ln \left(\frac{V}{V_o} \right)^n \tag{24}$$

which immediately suggests that $E/h\nu$ is discrete, an integer. That is, the total radiant energy is made up of discrete number of light quanta, each of energy $h\nu$. Einstein phrased it cautiously thus:

Light Quantum Hypothesis

Monochromatic radiation of low density (i.e. within the domain of validity of the Wien's radiation formula) behaves in thermodynamic aspect as if it consists of mutually independent energy quanta of magnitude $h\nu$.

The birth of quantum statistics two decades later brought to light so many non-rigorous features of Einstein's derivation. The analogy with a classical ideal gas is basically wrong, since photons do no obey Boltzmann statistics—they obey Bose statistics. Also, while the number of molecules in a closed system of an ideal gas is fixed, the photon number is not a conserved quantity. But luckily (for Einstein) the Boltzmann counting and Bose counting give the same answer in the Wien regime, while photon non-conservation hardly plays any role. This is why Einstein got away with the non-rigorous derivation and made the great discovery. As someone has said, for progress in science, one should know the truth but not the whole truth!

Photoelectric Effect

The light quantum hypothesis so far was just a formal result—an analogy. Now Einstein made a big jump and introduced what he called 'The Heuristic Principle'.

> If in regard to the volume dependence of entropy, monochromatic radiation (of sufficiently low density) behaves as a discrete medium of energy quantum of magnitude $h\nu$, then this suggests an inquiry as to whether the laws of generation and conversion of light are also constituted as if light were to consist of energy quanta of this kind.

The principle was applied in the 1905 paper to give a simple explanation of three experimental observations: frequency of emitted light cannot exceed that of incident light (Stokes' rule in photoluminescence); energy of electrons emitted in ionization of gases by ultraviolet light cannot exceed $h\nu$; and photoelectric effect.

Photoelectric effect discovered by Hertz in 1887 is the emission of electrons from metals when ultraviolet light is incident on them. Experiments by Lenard in 1902 showed that the maximum energy of the ejected electrons is independent of the intensity of radiation but increases with frequency. This observation could not be reconciled with the classical wave picture of radiation. In Einstein's picture, a photon of energy $h\nu$ transfers all its energy to a single electron, which leads to the Einstein's photoelectric equation

$$E_{\max} = h\nu - W \tag{25}$$

where W is the work function of the metal—the minimum energy required for an electron to come out of the metal. The equation was verified by an extensive series of experiments by Millikan around 1915. It also afforded another method, independent of Planck's Law, to determine h. The Heuristic Principle was later applied to Volta effect (inverse photoelectric effect), to determine the high frequency limit in Bremsstrahlung, etc.

The citation for the Nobel Prize in physics awarded to Einstein in 1922 referred to photoelectric effect but not specifically to his discovery of relativity.

> ... for his services to theoretical physics and in particular for his discovery of the law of the photoelectric effect.

Photoelectric effect was certainly a highlight of the application part of Einstein's March 1905 paper. Yet it was not photoelectric effect but rather the thermodynamics of Wien's radiation that led Einstein to the light quantum hypothesis.

Scepticism about Light Quanta

Unlike the discovery of special relativity which soon met acceptance and earned recognition for its creator, the light quantum hypothesis took almost two decades for its wide acceptance in physics. Einstein was himself quite tentative about it initially, as can be seen from the cautious way he expressed it in the March 1905 paper. As late as 1911, at the First Solvay Congress, he said thus: 'I insist on the provisional character of the concept (light-quanta), which does not seem reconcilable with the experimentally verified consequences of the wave theory.' Max Planck's reservation about the hypothesis is evident in his letter to Einstein in 1907. 'I am not seeking the meaning of the quantum of action (light quanta) in the vacuum but rather in places where absorption and emission occur and [I] assume that what happens in the vacuum is rigorously described by Maxwell's equations'. Planck's attitude to the March 1905 paper is best seen when he proposed Einstein for election to the Prussian Academy of Sciences:

> In sum, one can say that there is hardly one among the great problems in which modern physics is so rich to which Einstein has not made a remarkable contribution. That he may sometimes have missed the target in his speculations, as, for example, in his hypothesis of light quanta, cannot really be held too much against him, for it is not possible to introduce really new ideas even in the most exact sciences without sometimes taking a risk.

Above all, Millikan, who verified Einstein's photoelectric equation, better than anybody else at the time, did not accept his theory. 'Einstein's photoelectric equation ... appears in every case to predict exactly the observed results .... Yet the semi corpuscular theory by which Einstein arrived at this equation seems at present wholly untenable' (Millikan 1915).

What then turned the tables in favour of the light quanta? For Einstein, the turning point seemed to be his discovery in 1917 that the light quantum not only carries energy $h\nu$ but has a directed momentum $h\nu/c$. In a letter to his friend Besso in 1919, he said, 'I do not doubt any more the reality of radiation quanta, although I still stand alone in this conviction'. Experimentally, however, it was Compton Effect (1923) that seemed to have persuaded physicists to accept the notion of light quanta. Compton Effect is the scattering of a photon by a free electron. Applying energy and momentum conservation to photon-electron collision, we get the equation

$$\Delta\lambda = \frac{h}{mc}(1 - \cos\theta)$$

where $\Delta\lambda$ is the change in wavelength of the photon and θ the photon scattering angle. The verification of this equation indicated convincingly that 'a radiation quantum carries with it directed momentum as well as energy'. From then on light quanta were here to stay, though some people, notably Niels Bohr, continued to be sceptical much longer.

FROM LIGHT QUANTUM TO 'PHOTON'

The light quantum of the March 1905 paper was a parcel of energy. A photon (the term was coined by G.N. Lewis in 1926) carries an energy parcel of $h\nu$ and also has a directed momentum $h\nu/c$. It brings us much closer to viewing light as a stream of 'particles'. Einstein's solitary journey from light quantum to photon is an inspiring tale, full of gems of ideas that he collected on the way. The story becomes so much more inspiring when one remembers that another heroic saga of physics (creation of general relativity) was being written by the same man around the same period.

Statistical physics was Einstein's forte. He had in his early twenties arrived at several basics of this subject on his own, unaware that they had already been discovered, notably by Gibbs. In early 1909, Einstein applied his energy fluctuation formula

$$\langle \varepsilon^2 \rangle = \langle E^2 \rangle - \langle E \rangle^2 = kT^2 \frac{\partial \langle E \rangle}{\partial T} \tag{26}$$

to blackbody radiation between frequency ν and $\nu + d\nu$ and obtained the following result

$$\langle (\varepsilon^2(\nu, T) \rangle = \left(h\nu\rho + \frac{c^3}{8\pi\nu^2}\rho^2 \right) \nu d\nu \tag{27}$$

This formula is curious. If you used the classical Rayleigh-Jeans Law for radiation, you would get the second term above. On the other hand, the first term would result 'if radiation were to consist of independently moving point like quanta with energy $h\nu$'. This is a situation where the two opposing pictures of light (wave and particle pictures) do not appear in opposite limits but seem to be fusing together side by side in the same expression. This was the first indication of the latter day wave-particle duality.

In another work in 1909, Einstein considered momentum fluctuations in place of energy fluctuations. For a plane mirror of area f placed inside a cavity with radiation density ρ given by Planck's Law, he showed that the momentum fluctuation term Δ is given by

$$\langle \Delta^2 \rangle = \frac{1}{c} \left[\rho h\nu + \frac{c^3 \rho^2}{8\pi\nu^2} \right] f\tau d\nu \tag{28}$$

where τ is a small time interval and the mirror is taken to be fully transparent for all frequencies except those between ν and $\nu + d\nu$. Eq. (28) is closely analogous to eq. (27), with the second term corresponding to the wave picture (Rayleigh-Jeans Law) and the first term corresponding to the picture of 'point like quanta'. It is interesting that Einstein did not make a clear categorical statement based on the first term of eq. (28) that a light quantum of energy $h\nu$ has a momentum $h\nu/c$, though he did allude to the particle like aspect of light quanta in different ways.

Seven years later in 1916 (the long interlude being presumably due to his preoccupation with general relativity), came another beautiful work by Einstein, giving an alternative derivation of Planck's Law. Consider a molecular gas in thermal equilibrium with cavity radiation. The population N_n of the molecular energy level E_n is given by the Boltzmann factor

$$N_n = g_n e^{-E_n/kT} \qquad (29)$$

where g_n is the weight factor.

For two levels say E_1 and E_2 ($E_2 > E_1$), the transition rate between the levels is given by

$$R_{12} = N_1 \rho B_{12} \qquad (30)$$

$$R_{21} = N_2(\rho B_{21} + A_{21}) \qquad (31)$$

where R_{12} is the absorption rate, being naturally proportional to the population N_1 and the energy density of radiation. R_{21} is the emission rate. Here Einstein introduced the new concept of *stimulated emission*. R_{21} is a sum of two terms, one of which is proportional to radiation density ρ (stimulated emission) and the other independent of ρ (spontaneous emission). In equilibrium, the two rates are equal, which fact using eq. (29) gives:

$$A_{21}g_2 = \rho[B_{12}g_1 e^{(E_2-E_1)/kT} - B_{21}g_2] \qquad (32)$$

If we demand that eq. (32) yield Rayleigh-Jeans Law in the $T \to \infty$ limit, we get

$$B_{12}g_1 = B_{21}g_2 \qquad (33)$$

with the help of which eq. (32) may be rewritten as

$$\rho = \frac{A_{21}}{B_{21}}[e^{(E_2-E_1)/kT} - 1]^{-1} \qquad (34)$$

Planck's Law emerges if we put the Bohr condition

$$E_2 - E_1 = h\nu \qquad (35)$$

and

$$\frac{A_{21}}{B_{21}} = \frac{8\pi h\nu^3}{c^3} \qquad (36)$$

Eq. (36) could not be proved from first principles at that stage of physics. It was a consistency condition which together with Bohr condition and Boltzmann distribution formula guaranteed Planck's Law. It should be noted that the new concept of stimulated emission is necessary to arrive at Planck's Law. If this is not included, that is, the $B_{21}g_2$ term in eq. (32) is dropped; we will get Wien's Law, not Planck's Law. The practical demonstration of the notion of stimulated emission had to wait for the invention of masers and lasers some four decades later.

But Einstein did not stop at re-deriving Planck's Law. He went further. Molecules in equilibrium have the usual Maxwellian distribution of velocities.

But they are also incessantly under radiation pressure. How then is the Maxwellian distribution maintained in time? Einstein attacked the problem analogous to that of a mirror placed in a cavity. The details are involved, but the conclusion is striking. If a molecule absorbs a quantum of energy $h\nu$, a directed momentum $h\nu/c$ is transferred to it; likewise in emission, a momentum $h\nu/c$ is transferred to it opposite to the direction of the emitted light quantum. With this conclusion, the notion of light quantum with directed momentum $h\nu/c$ (photon) had been discovered in 1917, as far as Einstein was concerned. Its general acceptance had to wait till its experimental verification by Compton in 1923.

There was, however, one thing about the result that left Einstein uncomfortable. The molecule receives recoil in a certain direction during spontaneous emission of a light quantum. What determines the time when a photon is emitted spontaneously and what determines its direction? These questions intrigued Einstein and were the first hints clearly recognized by him of the breakdown of classical causality, which was to be the hallmark of modern quantum mechanics.

INDISTINGUISHABILITY OF LIGHT QUANTA

In 1924, the subject of light quanta witnessed an advance of great fundamental significance. In a letter to Einstein, a young Indian physicist, Satyendra Nath Bose, then a lecturer at Dacca University, gave an elegant new derivation of Planck's Law. Einstein was impressed by the paper; he translated the paper himself in German and forwarded it for publication in the *Zeitschrift fur Physik*. Much like Rayleigh's and unlike Planck's derivation and Einstein's 1916 derivation, Bose derived Planck's Law for a gas of photons without reference to any material particles in equilibrium with the radiation. In doing so, the essential (statistical) difference between the classical wave picture of radiation and the quantum picture of photons was laid bare. It took some time to distil out this difference from Bose's paper. When the difference became clear, the truly revolutionary nature of Bose's work emerged and the subject of quantum statistics was born.

In the classical picture, the number of electromagnetic standing waves with frequency ν to $\nu + d\nu$ per unit volume of cavity is given by $\dfrac{8\pi\nu^2 d\nu}{c^3}$. (Rayleigh-Jeans Law is obtained using this result and the classical equipartition theorem.)

Bose gave an incredibly simple derivation of the same result (using the 1917 result by Einstein that the momentum of a photon is $h\nu/c$) as follows: The total volume of one particle phase space with momentum between p and $p + dp$ is

$$V x 4\pi p^2 dp \qquad (37)$$

Multiply by 2 to count the two states of polarization to get

$$\text{total volume} = 8\pi V p^2 dp = \frac{8\pi V h^3}{c^3}\nu^2 d\nu \tag{38}$$

where $p = h\nu/c$ is used. Since each cell has volume h^3, the number of cells between ν and $\nu + d\nu$ is $\frac{8\pi V}{c^3}\nu^2 d\nu$, the same result as Rayleigh's.

Bose next pursued a counting procedure for calculating probability analogous to Boltzmann's but not quite. He counted in terms of cells not particles. Maximizing the probability, with total number of cells and total energy fixed as constraints, he arrived at Planck's Law. It took some time before it was finally realized that Bose's paper contained two fundamental (though related) ideas. First, photons are indistinguishable, in contrast to the distinguishability of, say, molecules of a classical ideal gas—the radiation state is defined by how many (not which) photons are there in different cells. Second, the different photons are not statistically independent. Einstein's phrase 'mutually independent light quanta' in his 1905 paper missed this point. There was also a third important idea—non-conservation of photons. It turns out that the constraint of 'number of cells fixed' in Bose's derivation is not really necessary: Planck's Law emerges with just 'total energy fixed' constraint.

Einstein soon extended Bose's derivation to the case of ideal gas of material particles, leading to the now well-known phenomenon of Bose-Einstein condensation—a demonstration of the quantum correlation in a gas of particles, even when they are non-interacting. In its pristine form, B-E condensation was experimentally seen in 1995. The quantum statistics followed by photons and material particles with integer spin (called Bosons) is known as Bose-Einstein statistics.

COMPLETING THE PICTURE BY DIRAC: END OF ACT ONE

The modern version of quantum mechanics came around 1925. It was important to obtain the B-E statistics in terms of the new mechanics by supplementing it with a detailed counting procedure. This task was accomplished by Dirac in 1926. Consider a classical ideal gas of N particles with total energy E. Suppose the single particle energy levels (discrete for simplicity) are denoted by ε_i, with population n_i. That is,

$$E = \sum n_i \varepsilon_i \quad N = \sum n_i \tag{39}$$

The number of microstates for this partition is given in Boltzmann statistics to be

$$W = \frac{N!}{\prod n_i!} \tag{40}$$

Dirac gave the prescription that for B-E statistics, the corresponding number W is just 1 (in view of indistinguishability of particles):

$$W = 1 \qquad (41)$$

and showed that it leads to Planck's Law. The associated N-particle quantum state is the unique completely symmetric state in N-particles. It turned out later that the N-particle quantum state of the so-called fermions (particles with half integer spin) is the unique completely anti-symmetric state. The quantum statistics obeyed by fermions is known as the Fermi-Dirac statistics.

Finally we return to Einstein's 1905 paper. It had none of the insights on photon statistics that came from Bose's paper, nor the ideas that came from Dirac in modern quantum mechanical terms. Yet it worked. Why? The most probable value $\langle n_i \rangle$ is given in Bose-Einstein statistics by

$$\langle n_i \rangle = \frac{1}{e^{h\nu_i/kT} - 1} \qquad (42)$$

In Wien's regime that Einstein had exploited, $h\nu_i/kT \gg 1$ that is $\langle n_i \rangle \ll 1$. In this limit, clearly, eqs (40) and (41) happen to be the same, except for the unimportant factor N! Einstein got away with his non-rigorous derivation because of this lucky circumstance!

In his famous 1926 work, Dirac quantized the electromagnetic field. The field was seen as a set of oscillators quantized as per the canonical procedure of the then new quantum mechanics. Photon was interpreted as an energy quantum of the electromagnetic field. The new discipline of Quantum Electrodynamics was born, which, among other things, validated the relation between the A and B coefficients of Einstein's 1916 paper. Act One of the story of light quanta was over.

Yet Einstein was not really comfortable with the notion of light quanta till the end of his life. Here was a revolutionary who doubted the basis of the revolution that he himself had helped trigger. Nowadays, with photon correlation experiments, which supposedly validate the quantum view, light quanta are part of the received wisdom in physics. But if Einstein's intuition is right, the final act of the story is perhaps still to be scripted.

REFERENCES

Mukunda, N. (2000). 'The Story of the Photon', *Resonance*, 5(3), 35.

Pais, Abraham (1982). *Subtle is the Lord: The Science and the Life of Albert Einstein*. London and New York: Oxford University Press.

Richtmyer, F.K., E.H. Kennard, and J.N. Cooper (1969). *Introduction to Modern Physics* (6th edn.), New York: McGraw-Hill.

Stachel, John (ed.) (1998). *Einstein's Miraculous Year*. Princeton, NJ: Princeton University Press.

3

JAYANT VISHNU NARLIKAR

Einstein and Cosmology

HISTORICAL BACKGROUND

The year 1905 is a historical year for physics and for Albert Einstein. It was the year when Einstein published his three papers on fundamental character (Einstein 1905a, b, c), on Brownian motion, photoelectric effect, and the special theory of relativity. The first gave a statistical description of molecular motion in fluids and enriched our understanding of the microscopic composition of fluids. The second was the initiator of the idea that light can also be described as made of particles subject to the discipline of quantum mechanics, while the third led to a revision of the very basic ideas of space-time measurements. It is but natural that we celebrated the international year of physics as a centenary of the above developments in 2005.

Ten years after proposing his special theory of relativity, Einstein came out with a more comprehensive general theory of relativity, which also provided a very unusual description of the phenomenon of gravity as a manifestation of curved space-time around any presence of matter and energy. After coming out with the general theory in 1915, Albert Einstein (1917) used it in an ambitious way to propose a model of the entire universe. This simple model assumed that the universe is homogeneous and isotropic and *also static*. Homogeneity means that the large-scale view of the universe and its physical properties at any given epoch would be the same at all spatial locations. Isotropy demands that the universe looks the same in all directions, when viewed from any spatial location. The requirement of a static universe was motivated by the then perception that there is no large-scale systematic movement in the universe.

That was the general belief at the time. In fact the realization that there is a vast world of galaxies spread beyond the Milky Way had not yet seeped into the astronomical community. Although there were isolated measurements of nebular redshifts, these did not convey any impression that the universe as a whole is not static. However, to obtain such a static model Einstein had to modify his general relativistic field equations to include an additional *cosmological constant term* λ which corresponded to a long-range force of repulsion.

Since this is a non-technical account, I am in general staying away from formulae and equations. However, in this particular instance, I feel that I must state the modification made by Einstein in technical jargon. I will, however, point out the change alone, without describing the mathematical details. I think the change made by Einstein can be understood without technical details.

The original equations were:

$$R_{ik} - 1/2 g_{ik} R = -[8\pi G/c^4] T_{ik} \qquad (1)$$

I have used the common notation in which G stands for the Newtonian constant of gravitation and c stands for the speed of light. These are two fundamental constants of nature.

Here the left-hand side relates to the space-time geometry of the universe and the right-hand side describes the physical contents of the universe. These equations, however, could not yield a static model of the universe as a solution and so Einstein sought to modify them in the *simplest possible way*. This led him to the following equations:

$$R_{ik} - 1/2 g_{ik} R + \lambda g_{ik} = -[8\pi G/c^4] T_{ik} \qquad (2)$$

Notice the extra term on the left-hand side, which has introduced a new constant of nature, λ. In the 'Newtonian approximation' this additional term corresponds to an acceleration of $\lambda r c^2$ between any two-matter particles separated by a distance r. The term λ is called the cosmological constant since its value is very small (today's estimate is $\sim 10^{-56}$ cm^{-2}), and it does not affect the motion of matter significantly on any but the cosmological scale. Thus, to all intents and purposes, the gravitational effects on the scale of the solar system or the stars and galaxies remain unaffected.

The Einstein universe, as the model came to be known, described the universe as a three-dimensional surface of a four-dimensional 'hypersphere', which does not change with time. The Einstein universe is homogeneous and isotropic, that is, at any time the universe looks *the same* from any vantage point and also in any direction. The field equations (2) then give the density and radius of the universe in terms of the fundamental constants G, c, and λ. To Einstein this was an eminently satisfactory outcome as it related physics of the universe to its space-time geometry in a unique way. The gravity of the matter 'curled up' the space into a finite volume, showing the essence of the general relativistic relationship between gravity and space curvature. He felt that the uniqueness of the solution attached special significance to the model in terms of credibility.

He was in for disappointment on this count as within a few months de Sitter (1917) found another solution completely different from Einstein's. The de Sitter universe was empty and *non-static*. The space in the de Sitter universe shows continual steady expansion. One can say that whereas the Einstein universe had matter without motion, the de Sitter universe had motion without matter! In 1917, the astronomical data did not support the de Sitter model, which therefore remained a mathematical curiosity.

In 1922–4, Alexander Friedmann, however, showed that one can obtain homogeneous and isotropic solutions without the cosmological term, but they describe models of an *expanding universe* (Friedmann 1924). In 1927, Abbé Lemaître also obtained similar solutions, but these, along with the Friedmann models were considered as mathematical curiosities (Lemaître 1927). The Friedmann-Lemaître models thus shared the feature exhibited by the de Sitter universe, namely that of expansion.

Meanwhile, on the observational side, the early (pre-1920) perception of a universe mostly confined to the Milky Way Galaxy with the Sun at its centre, eventually gave way to the present extra-galactic universe in which our location has no special significance. Indeed this 1905 quotation of Agnes Clerke in her popular book on astronomy expresses the prevalent dogma of those times.

The question whether nebulae are external galaxies hardly any longer needs discussion. It has been answered by the progress of research. No competent thinker, with the whole of the available evidence before him, can now, it is safe to say, maintain any single nebula to be a star system of co-ordinate rank with the Milky Way. A practical certainty has been attained that the entire contents, stellar and nebula, of the sphere belong to one mighty aggregation, and stand in ordered mutual relations within the limits of one all embracing scheme.

This perception represented the majority view, which was still existent in 1920, when the famous Shapley-Curtis debate took place. Shapley spoke in support of this view while Curtis represented the slowly emerging view that many of the faint nebulae were external galaxies far away from the Milky Way.

During the 1920s Edwin Hubble gradually established this picture in which spiral and elliptical galaxies are found all over the universe. The erroneous observations of Adriaan van Maanen contradicting this picture and arguing that all spiral nebulae were galactic had been influential in the delay in accepting this revised picture. These were eventually set aside. In 1929, Hubble established what is today known as the Hubble Law which is generally interpreted as coming from an expanding universe (Hubble 1929). In this, Hubble spectroscopically determined the Doppler radial velocities of galaxies and found these to vary in proportion to their distances.

Typically, one finds that a standard dark line in the spectrum of a galaxy does not appear at its laboratory wavelength, but at longer wavelength. The fractional increase in the wavelength is called 'redshift' and denoted by the symbol z. (The name is because the spectrum appears to have shifted to the red-end of the seven-colour spectrum.) For example, if the laboratory-measured wavelength of the line was 400 nanometres and it appears with a wavelength of 500 nanometres, then for this line $z = 0.25$. The well-known Doppler Effect relates such redshifts to the radial velocity v of the source of light away from the observer. A redshift of 0.25 would mean that the source galaxy is moving away from us with a speed $0.25c$. Thus one may write Hubble's Law in terms of redshifts as:

$$v \equiv c \times z = H \times D \qquad (3)$$

where D is the distance of the extragalactic object with redshift z. The constant of proportionality is called the Hubble constant and it is commonly denoted by H. (Caution: We have used the Doppler formula for speeds small compared to c.)

The Friedmann-Lemaître models now no longer were mathematical curiosities, but were seen as the correct models to explain Hubble's Law. A typical model in this class is described by a time-dependent scale factor S which is increasing with time at present, and a curvature parameter k, which indicates if the space is flat ($k = 0$), positively curved like the surface of a sphere ($k = 1$), or negatively curved like the surface of a saddle ($k = -1$). The Einstein universe had $k = 1$ whereas the de Sitter universe had $k = 0$. The motion of the galaxy is manifest through the changing scale factor $S(t)$. The redshift is interpreted in terms of this model as coming from a time-dependent scale-factor $S(t)$: if the light signal from the source left at time t_1 and it reached the observer at time t_0 then we have

$$1 + z = S(t_0)/S(t_1) \qquad (4)$$

The scale-factor $S(t)$ and the curvature parameter k were to be determined from Einstein's field equations. Einstein also decided that his cosmological constant was no longer needed and gave it up. Incidentally the much-publicized remark by Einstein that the cosmological constant was the 'greatest blunder' of his life has no direct authentication in Einstein-literature. It has been ascribed to George Gamow who claimed that this is what Einstein said to him (Gamow 1971).

The stage was thus set to launch cosmology as a discipline wherein the theoretical predictions based on relativistic models could be tested by observations of the extragalactic universe.

EARLY COSMOLOGY

During the 1930s, cosmologists led by Eddington (1930) and Lemaître (1931) discussed the theoretical models of the expanding universe and all these led to the concept of a 'beginning' when the universe was dense and very violent. Lemaître called the state that of a *primeval atom*. Later, Fred Hoyle, an opponent of this idea referred to the state as of 'big bang', a name that caught on when the model became more popular.

The crucial effect in Hubble's Law was the redshift found in the spectra of galaxies and its progressive increase with the galactic distances. The linear law discovered by Hubble was believed to be an approximation of the exact functional relationship between redshift and distance according to any of the various Friedmann-Lemaître models. Attempts were made by succeeding astronomers to carry out deeper surveys to test the validity of this extrapolation.

Hubble's own priorities on the observational side were elsewhere (1938). He wanted to fix the value of the mathematical parameter k of the model by observing galaxies and counting them to larger and larger distances. He made several unsuccessful attempts before realizing that the ability of the 100-inch Hooker telescope fell short of making a significant test of the relativistic models. He proposed the 5-metre telescope at the Palomar Mountain for this very reason as this bigger telescope was expected to settle this cosmological problem. By the time the telescope was completed and began to function (late 1940s) Hubble had realized that his observational programme was not a realistic one and the telescope, in fact, came to be used for other important works.

The reason Hubble's programme was unworkable was that in order to detect the effects of space-time curvature through galaxy counts, one needed to look very far out, to redshifts of the order of unity, and this requirement was hard to satisfy for two reasons.

1. Observational techniques were not yet sophisticated enough to detect galaxies of such large redshifts.
2. The number of galaxies to be counted was enormously large if one were to use the counts to be sensitive enough to draw cosmological conclusions. There was a third difficulty with the number count programme, to which I shall return later.

Meanwhile, we should now take note of another major observational front opening in astronomy, which till then had been dominated by the monopoly of optical astronomy.

THE ADVENT OF RADIO ASTRONOMY

Astronomy became more versatile after the Second World War, when radio astronomy came into existence as a viable tool of observation. In their enthusiasm about the new technique, radio astronomers felt that they could undertake Hubble's abandoned programme by applying it to the counts of radio sources. In the 1950s radio astronomers in Cambridge, England, and in Sydney as well as Parkes, Australia, began their attempts to solve this problem by counting radio sources out to very faint limits. Radio astronomy apparently got round the two difficulties mentioned above. Radio galaxies could be observed, it was felt, to greater distances than optical galaxies and there were far fewer of them to count.

The basic test of counting of radio sources went thus. If one accepts that radio sources are of uniform luminosity and are homogeneously distributed in the universe, then in the static Euclidean model, it can be easily shown that the number (N) – flux density (P) satisfies the relation

$$\log N = -1.5 \log P + \text{constant} \qquad (5)$$

The relation for a typical expanding Friedmann-Lemaître model shows that at high densities the number count N rises with diminishing flux according to relation (5) but at lower flux densities it rises slowly than shown by (5). However, if one puts in an ad hoc assumption that the number density of radio sources per unit co-moving coordinate volume was higher than at present, then one can get slopes steeper than -1.5.

While the Australians felt that within the existing error-bars, their surveys did not show any evidence inconsistent with the Euclidean model, the Cambridge group under the leadership of Martin Ryle made several claims to have found a steep slope. While the early Cambridge data were later discounted as being of dubious accuracy, the data in the early 1960s (the 3C and 4C surveys) did show a slope of -1.8 at high flux density, which subsequently flattened at low flux densities. The steepness was claimed by Ryle to have confirmed the big bang models. However, it later became clear that these radio surveys might tell us more about (1) local inhomogeneity and (2) the physical properties of the sources rather than about the large-scale geometry of the universe (Narlikar 2002).

THE STEADY STATE THEORY

In 1948, a rival to the classic big bang theory emerged. This theory was based on a model of the universe with the de Sitter metric, but which had a constant non-zero density of matter (Bondi and Gold 1948; Hoyle 1948). Such a model can be obtained from Einstein's gravitational equations (without the cosmological term), provided on the right-hand side one introduces a negative energy field, called originally the C-field. Hoyle and later Maurice Pryce (private communication) worked on the C-field concept and a theory based on a scalar field derivable from an action principle emerged in 1960. This idea was developed further by Hoyle and Narlikar (1966). Although the concept of a negative energy scalar field was considered by physicists to be unrealistic in the 1960s, today, four decades later it is appreciated that the currently popular phantom fields are no different from the C-field.

Since, as the name implies, the steady state theory described an unchanging universe (on a large enough scale), the observational predictions of the theory were unambiguous and this was cited as a strength of the theory. Ryle's main attack was directed against this theory with the assertion that the radio source counts disproved this theory. This claim was refuted by Hoyle and Narlikar (1961) with the demonstration that in a more realistic structure of the universe inhomogeneities on the scale of 50–100 Mpc (megaparsec: 1

parsec is approximately 3 light years) would give rise to steep slopes of the log N – log P curve for radio sources.

Although the steady state theory survived Ryle's challenges, it appeared to receive a mortal blow in 1965 by the discovery of the cosmic microwave background. Also, it could not account for the rather large fraction (~ 25 per cent) by mass of helium in the universe. To understand the implications of this result one needs to look back at the studies of the early universe in relativistic cosmology.

THE EARLY HOT UNIVERSE

In the mid-1940s, George Gamow (1946, 1948) started a new programme of studying the physics of the big bang universe close to the big bang epoch. For example, calculations showed that the universe in its early epochs was dominated by relativistically moving matter and radiation and that the temperature T of the universe, infinite at the big bang, dropped according to the law:

$$T = B/S \cdot B = \text{constant} \qquad (6)$$

As the universe expands, its temperature drops, just as a volume of hot gas cools down as it expands. Thus the temperature of the universe fell to about ten thousand million degrees after one second. In the era 1-200 second, Gamow expected thermonuclear reactions to play a major role in bringing about a synthesis of the free neutrons and protons that were lying all over the universe. Were all the chemical elements we see today in the universe formed in this era?

This expectation of Gamow turned out to be incorrect. Only light nuclei, mainly helium could have formed this way. Also, one could adjust the density of matter in the universe over a wide band to produce the right cosmic abundance of helium. The heavier elements could, however, be formed in stars, as was shown later by the comprehensive work of Geoffrey and Margaret Burbidge, William Fowler, and Fred Hoyle (1957). Today it looks as if the light nuclei were made in Gamow's early universe, as the stars do not seem to be able to produce them in the right abundance. It was because of this circumstance that the steady state universe, which did not have a very hot era, failed in the production of helium.

Apart from this evidence, there was another prediction made by Gamow's younger colleagues, Ralph Alpher and Robert Herman (1948), namely that the radiation surviving from that early hot era should be seen today as a smooth Planckian background of temperature of around 5 K. This prediction has been substantiated. In fact in 1941, McKeller (1941) had deduced the existence of such a background of temperature 2.3 K from spectroscopic observations of CN and other molecules in the Galaxy. This result was not widely known or appreciated at the time. In fact it was the serendipitous observation of an isotropic radiation by Arno Penzias and

Robert Wilson (1965), that drew physicists and cosmologists to the big bang model in a big way. Penzias and Wilson found the temperature to be 3.5K.

The post-1965 development of cosmology took a different turn. The finding of the cosmic microwave background radiation (CMBR) was taken as vindication of the early hot universe and efforts were made to observe the spectrum of the radiation as accurately as possible. In 1990, the COBE satellite gave a very accurate Planckian spectrum thus providing confirmation of the Alpher-Herman expectation of a relic black body spectrum (Mather 1990). Another expectation, of finding small-scale inhomogeneities in the background was also fulfilled two years later when COBE found (Smoot 1992) such fluctuations of temperatures $\Delta T/T$ of the order of a few parts in a million. On the theoretical side the emphasis shifted from general relativistic models to models of a very small-scale universe with high temperature corresponding to fast-moving particles. Theorists also began to come to grips with the problem of formation of large-scale structures ranging from galaxies to superclusters. We will consider these developments next.

PHYSICS OF THE EARLY AND VERY EARLY UNIVERSE

The CMBR prompted many physicists to look in depth at the physics of the post- and pre-nucleosynthesis era. For example, as the universe cools down, the chemical binding can become important and trap the free electrons into protons to make neutral hydrogen atoms. This eliminates the major scattering agency from the universe and radiation can subsequently travel freely. Calculations show that this epoch was at redshift of around 1000–1100 (Weinberg 1972).

If instead we explore epochs *earlier* than the nucleosynthesis one, we would encounter higher temperature and more energetic activity. This has attracted particle physicists to the big bang models for here they have a possibility of testing their very high-energy physics. The very early epochs when the universe was 10^{-38} second old had particles of energy so high that they might have been subject to the grand unification scheme, which could therefore be tested. Energies required for such testing are, however, some 13 orders of magnitude higher than what can be produced by the most powerful accelerators on the earth.

Such a combination of disciplines is called Astroparticle Physics. One of its most influential 'gifts' has been the notion of inflation (Kazanas 1980; Guth 1981; Sato 1981). This is the rapid exponential expansion of the universe lasting for a very short time, produced by the phase transition that took place when the grand unified interaction split into its component interactions (the strong and electroweak interactions). Inflation is believed to solve some of the outstanding problems of the standard big bang cosmology, such as the horizon problem, the flatness problem, the entropy problem, etc.

DARK MATTER AND DARK ENERGY

One of the conclusions of inflation is that the space part of the universe is flat. Theoretically it requires the matter density to be $\rho_c = 3H^2/8\pi G$. Here H is the Hubble constant and G is the gravitational constant. This value, sometimes known as the *closure density*, leads straightaway to a conflict with primordial nucleosynthesis, which tells us that at this density there would be almost no deuterium produced. Even if we ignore inflation, and simply concentrate on the empirical value of matter density determined by observations, we still might run into a serious conflict between theory and observation: there is evidence for greater matter density than permitted by the above deuterium constraint.

For, while the visible matter in the form of galaxies and intergalactic medium leads to a value of density which is less than 4 per cent of the closure density, there are strong indications that additional *dark* matter may be present too (Narlikar 2002). The adjective 'dark' indicates the fact that this matter is unseen but exerts gravitational attraction on visible matter. Such evidence is found in the motions of neutral hydrogen clouds around spiral galaxies and in the motions of galaxies in clusters. Even this excess matter would cause problem with deuterium.

To get round this difficulty, the big bang cosmologists have hypothesized that the bulk of dark matter is *non-baryonic*, that is, it does not influence nucleosynthesis. Writing the ratio of the density of non-baryonic matter to the closure density as Ω_{nb} and the corresponding ratio for baryonic matter as Ω_b, we should get as per inflation $\Omega_{nb} + \Omega_b = 1$. Thus, if the baryonic matter is 4 per cent, the non-baryonic matter should be 96 per cent.

However, even this idea runs into difficulty, as there is no direct evidence for so much dark matter. A solution is provided, however, by resurrecting the cosmological constant that Einstein had abandoned in the 1930s. We can define its relative contribution to the dynamics of expansion through a parameter analogous to the density parameter:

$$\Omega_\Lambda = 3\lambda H^2/c^2 \qquad (7)$$

Thus we now get something like: $\Omega_b = 0.04$, $\Omega_{nb} = 0.23$, and $\Omega_\Lambda = 0.73$. So, the extra energy put in is called *dark energy*. The total of these values is meant to add up to unity, as expected by the inflationary hypothesis.

STRUCTURE FORMATION

These issues are important to the understanding of how large-scale structure developed in the universe. To this end, the present attempts assume that small fluctuations were present in the very early universe and these grew because of inflation and subsequent gravitational clustering. Various algorithms exist for developing this scenario. One of the basic inputs is the way the total density is split up between baryonic matter, non-baryonic matter, and dark energy.

The non-baryonic dark matter can be hot (HDM) or cold (CDM) depending on whether it was moving relativistically or non-relativistically at the time it decoupled from ordinary (baryonic) matter.

A constraint to be satisfied by this scenario is to reproduce the observed disturbances generated in the CMBR by these agents and also the observed extent of clustering of galaxies today. For, observations of small inhomogeneities of the CMBR rule out various combinations and also suggest what kind of dark matter (cold or hot or mixed) might be required. Currently the model favoured is called the Λ CDM-model to indicate that it has dark energy and cold dark matter.

OBSERVATIONAL TESTS

Like any physical theory cosmology also must rely on observational tests and constraints. There are several of these. There have been tests of cosmological models of the following kinds:

1. Geometry of the universe
2. Physics of the universe

The first category includes the measurement of Hubble's constant, the redshift (z), and magnitude (m) relation to high redshifts, the counting of radio sources and galaxies, the variation of angular size with redshift, and the variation of surface brightness with redshift. The apparent magnitude is a measure of the brightness of the source on a logarithmic scale. For a family of sources of the same luminosity, m is thus a measure of distance of a source ... the larger the value of m the more distant is the source.

The measurement of Hubble's constant has been a tricky exercise right from the early days dating back to Hubble's original work. The problem is to be sure that no systematic errors have crept in the distance measurement, as these have not yet been fully debugged. Which is why we still have serious observing programmes yielding values close to 70 km/s/Mpc as well as to 55 km/s/Mpc. At the time of writing this review, the majority opinion favours the higher value, but 'rule of the majority' has not always been a successful criterion in cosmology.

The measurement of the z-m relation had been attempted by Allan Sandage for quite a long time and during the period 1960–90 the overall view was that the relation as applied to the brightest galaxies in clusters treated as standard candles, favoured *decelerating* models. These models are naturally given by the Friedmann solutions *without* the cosmological constant. However, in the late 1990s, the use of Type 1a supernovae has led to a major reversal of perception and the current belief is that the universe is *accelerating* (Reiss et al. 1998; Perlmetter 1999). The other tests like number counts or angular size variation have not been so clear-cut in their verdict as they get mixed up with evolutionary parameters. Apart from the difficulties encountered by Hubble in the 1930s, any cosmological test using source populations of a certain type necessarily gets involved with the possibility that the source yardstick may be evolving with age.

Currently, cosmologists are most attracted to measurements of the angular power spectrum of the microwave background inhomogeneities. These can be related to other dynamical features of the universe, given a cosmological model satisfying Einstein's equations with the cosmological constant. Using details from the WMAP satellite one can get a range of models with $k = 0$. Among these models those with a positive cosmological constant are favoured. As mentioned before, the favoured solution has $\Omega_b = 0.04$, $\Omega_{nb} = 0.23$, and $\Omega_\Lambda = 0.73$. We recall that the low value of baryonic density is required to understand the abundance of deuterium.

Many cosmologists feel that, there, is now a 'concordance' between various tests that suggest the above combination for the energy content of the universe together with the higher of the two values of the Hubble constant mentioned above. It is felt that this set of parameters describes accurately most of the observed features of the universe. With this optimistic view one may be tempted to think that the quest for the model of the universe that began with Einstein in 1917 is coming to an end.

NEED FOR CAUTION AND ALTERNATIVES

However, there needs to be some caution towards this optimism. The concordance has been achieved at the expense of bringing in a lot of speculative element into cosmology. Thus, there is as yet no independent evidence for the non-baryonic dark matter, nor any for the dark energy. The notion of inflation is widely believed in, but as yet there is no physical theory for it within the overall framework of high energy physics, nor is the era of the inflation directly observable by telescope. Thus, we are asked to believe in speculative idea with no direct theoretical framework or observational validation. Then a lot revolves round the concept of inflation, which is still not describable as a process based on a firm physical theory. The densities of matter one is talking about when inflation took place were some 10^{50} times the density of water. Recall how much investigation went into the equation of state for neutron stars where the matter density was a mere 10^{15} times the density of water. Yet we do not find any discussion of how such matter behaves in real life. Likewise, the inflationary time scales of the order of 10^{-38} second defy any operational physical meaning. These are some 25 orders of magnitude smaller than the shortest measurable time scale known to physics, viz. those measured by the atomic clocks. So a physicist may wonder if the concordance cosmology is a rigorous physical exercise at all.

Today the concordance picture looks good if one is happy with the number of epicycles that have gone into it. Non-baryonic dark matter and dark energy are two of them. They had to be introduced in order to ensure the survival of the model: they have no independent direct confirmation. These are examples of extrapolations of known physics to epochs that are astronomically unobservable. While indirect observations showing an overall consistency of these assumptions are necessary for the viability of the concordance model, they cannot be considered sufficient.

This is why there appears to be need for new ideas in cosmology, especially alternative scenarios that are less speculative and follow very different tracks from the above standard scenario. Some attempts are in vogue at present, like the Quasi-Steady State Cosmology (QSSC) (Hoyle et al. 2000) or the Modified Newtonian Dynamics (MOND)(Milgrom 1983), which are, however, very much minority efforts. Perhaps by 2017, a hundred years after Einstein's paper on cosmology we may have a more realistic perception of how complex our universe is. I can do no better than end with a quotation from Fred Hoyle:

> ... I think it is very unlikely that a creature evolving on this planet, the human being, is likely to possess a brain that is fully capable of understanding physics in its totality. I think this is inherently improbable in the first place, but even if it should be so, it is surely wildly improbable that this situation should just have been reached in the year 1970 ... (1970)

Fred Hoyle said this at the Vatican Conference held towards the end of the 1960–70 when cosmologists were making equally confident remarks about how well the universe was being understood. This was before inflation, dark matter, dark energy, etc. were even thought of. Are today's cosmologists sure that they have all pieces of the jigsaw puzzle that make up our universe?

REFERENCES

Alpher, R.A., H.A. Bethe, and G. Gamow (1948). *Phys. Rev.*, 73, 80.

Alpher, R.A. and R.C. Herman (1948). *Nature*, 162, 774.

Bondi, H. and T. Gold (1948). *MNRAS*, 108, 252.

Burbidge, E.M., G.R. Burbidge, W.A. Fowler, and F. Hoyle (1957). *Rev. Mod. Phys.*, 29, 547.

de Sitter, W. (1917). *Koninkl. Akad. Weteusch Amsterdam*, 19, 1217.

Einstein, A. (1905a). *Annalen der Physik*, ser. 4, 17, 132.

——— (1905b). *Annalen der Physik*, ser. 4, 17, 549.

——— (1905c). *Annalen der Physik*, ser. 4, 17, 891.

——— (1917). *Preuss. Akad. Wiss. Berlin Sitzber*, 142.

Eddington (1930). A.S., *MNRAS*, 90, 668.

Friedmann, A. (1922). *Z. Phys.*, 10, 377.

——— (1924). *Z. Phys.*, 21, 326.

Gamow, G. (1946). *Phys. Rev.*, 70, 572.

——— (1971). *My World Line*, Viking Adult.

Guth, A.H. (1980). *Phys. Rev. D*, 23, 347.

Hoyle, F. (1948). *MNRAS*, 108, 372.

———— (1970). *Study Week on Nuclei of Galaxies*, in D.J.K. O'Connell (ed.), North Holland, Amsterdam, p. 757.

Hoyle, F. and J.V. Narlikar (1961). *MNRAS*.

———— (1962). *Proc. Roy. Soc.* A270, 334.

———— (1966). *Proc. Roy. Soc.* A290, 162.

Hoyle, F. G. Burbidge, and J.V. Narlikar (2000). *A Different Approach to Cosmology*, Cambridge.

Hubble, E.P. (1929). *Proc. Nat. Acad., USA*, 15, 168.

———— (1938). *Ap.J.*, 84, 517.

Kazanas, D. (1980). *Astrophysical Journal*, 241, L59.

Lemaître, Abbé G. (1927). *Annales de la Société Scientifique de Bruxelles*, XLVIIA, 49.

———— (1931). *MNRAS*, 91, 490.

McKellar, A. (1941). *Pub. Dom. Astrophys. Obs. Victoria*, 7, 251.

Milgrom, M. (1983). *Astrophysical Journal*, 271, 365, 371, and 383.

Narlikar, J.V. (2002). *An Introduction to Cosmology* (3rd edn.), Cambridge.

Penzias, A.A. and R.W. Wilson (1965). *Astrophysical Journal*, 142, 419.

Perlmutter S. et al. (1999). *Astrophysical Journal*, 517, 565.

Reiss, A. et al. (1998). *Astrophysical Journal*, 116, 1009.

Sato, K. (1981). *MNRAS*, 195, 467.

Smoot, G. et al. (1992). *Astrophysical Journal*, 396, L1.

Spergel, D.N. et al. (2003). *Astrophysical Journal, Suppl.*, 148, 175.

Weinberg, S. (1972). *Gravitation and Cosmology*, John Wiley, New York.

4

SHASHIKUMAR MADHUSUDAN CHITRE

Role of Relativity in Astronomy and Astrophysics

Stars and galaxies constitute the main building blocks of the universe. In the first half of the twentieth century it was the structure and thermonuclear evolution of stars, while in the second half it was the formation and chemical evolution of galaxies that dominated the course of astrophysical research. Despite the pre-eminent position of Einstein in the realm of physics, general relativity had hardly any role to play in astrophysics during this period. This was understandable because general relativistic effects associated with the gravitational bending of light and the precession of peribelion of planet Mercury's orbit was demonstrably tiny to be of any relevance in the wider domain of astrophysics.

In the realm of extragalactic astronomy, once the expansion of the universe was observationally established by Hubble (1934), general relativity continued to direct the course of observational cosmology. During this period an important avenue of research was initiated for outlining the thermodynamic history of the universe and for predicting the possible observable remnant of the very early hot, dense phase of an expanding universe. It also opened up the tantalizing possibility of manufacturing the light elements up to lithium within the first three minutes after the creation of the cosmos. A major contribution of nuclear astrophysics was to establish that the elements heavier than lithium are all practically produced inside massive stars (Burbidge et al. 1957). It turns out the evolution of atoms in the universe is closely linked with the evolution of stars.

Special relativity, however, had played a crucial role in solving the central problem concerning the source of energy for stars. The process of thermonuclear fusion is a direct consequence of Einstein's famous expression for mass and energy conversion, $E = mc^2$ where m is the rest mass. Thus, the proton-proton chain and carbon-nitrogen-oxygen cycle in which hydrogen is converted into helium in order to provide the source of nuclear energy required to sustain the luminosity of the star over hundreds of millions billions of years, since the combined mass of nucleons exceeds the mass of the helium nucleus. Likewise, the principle of time-dilation is the basis for being able to detect the penetrating component of cosmic rays at mountain altitude and sea-level though they are produced through the decay of pi-meson produced in the collisions of primary cosmic rays at high altitudes—typically 20 km in the atmosphere. The decay time of the μ meson is 2 microseconds in its rest frame and can travel only 0.2 km. The relativistic time dilation stretches the lifetime by a factor of 100 and the moon is able to travel 20 km before decaying. In fact this was the first experimental evidence for time dilation. Special relativity has also had its impact in the field of spectroscopy with special reference to Doppler broadening of special life and the interaction of radiation (photon) with matter. It also elucidated the pattern of synchrotron radiation emitted by non-thermal relativistic motion of charged particles (electrons) in the presence of magnetic fields, which forms the basis for radio emission from extragalactic sources. The fruitful confluence of special relativity and quantum physics, of course, resulted in the prediction of the limiting mass, for white dwarfs (Chandrasekhar 1935), thus illuminating the end-point of evolution of massive stars in the form of neutron stars and stellar mass black holes.

PHYSICS AND ASTROPHYSICS OF CONDENSED OBJECTS

Nature of Collapsed Objects

A tenuous interstellar gas cloud composed largely of hydrogen and helium can become gravitationally unstable and condense to form a star, which is constantly battling against the force of gravity. The resulting contraction of the protostar leads to an increase in the central density and temperature in accordance with,

$$\rho \approx M/R^3 \qquad (1)$$

$$kBT \approx mpGM/R \qquad (2)$$

Here M is the mass and R, the radius of the quasi-spherical star, kB is the Boltzmann constant, G, the gravitational constant, and mp, the proton mass. The life history of a star is characterized by a continued contraction of its central regions with halts caused by successive nuclear energy generation processes. These progressively go through the burning of hydrogen, helium, carbon, oxygen, neon, and silicon fuels until the nuclear statistical peak is reached around iron. Of course, the full sequence of nuclear burning is followed only by the massive stars, while in low-mass ($M \approx 0.08 M\odot$) stars, the central temperature never attains a high enough value ($T > 10^7 k$) to ignite the nuclear fuel for overcoming the Coulomb barrier to fuse the protons.

However, in the process of contraction the interior could become sufficiently dense for the degeneracy pressure of electrons to counter the gravitational force. These stars known as brown dwarfs are never able to energize themselves at any point of their long uneventful existence (Kulkarni 1997).

The temperatures inside stars like our sun reach values upwards of 1.5×10^7 deg. K when the hydrogen in the core burns to form helium, mainly by the proton-proton nuclear reaction chain. Once the hydrogen in the central region is exhausted, the core contracts, raising its temperature to some 10^8 K when other high-temperature reactions set in, converting helium into carbon and oxygen. A star, which initially has a mass approximately between one to eight solar masses, develops a carbon-oxygen core with the hydrogen and helium burning shells surrounding it. With the burning of these shell sources towards the surface, the star becomes progressively more luminous and eventually the envelope is blown away by the action of powerful radiation pressure and fast stellar winds. The mechanism of mass loss and the processes arising from the combined effect of rotation and magnetic field lead to a variety of planetary nebular morphologies, in the shape of rings, bipolar or butterfly configurations (Weinberger and Kerber 1997). The carbon-oxygen cores left behind become white dwarfs which have masses comparable to that of the sun, but their sizes are more like that of earth. The central densities are in the range of $10^6 - 10^7$ gm/cm^3 and there is a limit to the mass of the white dwarf that can be supported by the pressure provided by the degenerate electrons. This limiting mass for white dwarfs, is the famous Chandrasekhar limit (1935),

$$Mch \approx (hc/Gm_p^2)^{3/2} m_p \approx 1.4 \, M\odot \tag{3}$$

a brilliant demonstration of the confluence of special relativity and quantum mechanics with the assistance of gravity.

Following the development of carbon-oxygen core, there is a branching point and stellar evolutionary calculations suggest that if the core is not sufficiently massive this thermonuclear evolutionary sequence can stop at this point. However, stars with initial masses larger than 8 M⊙ can undergo further core collapse and can ignite successive nuclear fuels. The star will then assume an onion-like structure, composed of an iron-nickel core surrounded by concentric shells of silicon, magnesium, oxygen, neon, carbon, helium, and an envelope of unprocessed hydrogen. The iron-nickel core becomes unstable because of the capture of electrons by nuclei and also by their photodisintegration. This removes the main pressure-support of the outer layers leading to a collapse of the star and within a few seconds the imploding core gets detached from the envelope. When the collapsed core attains a density of the order of nuclear density ($\approx 2 \times 10^{14}$ gm/cm^3), it resembles a giant nucleus made up of neutrons, protons, and electrons. The degenerate pressure provided by neutrons and the strong repulsive nuclear forces then come into play in order to stall the collapse. Such a sudden halting of the infalling matter induces a conversion of the enormous gravitational energy released ($\sim 10^{53}$ ergs) into thermal energy, generating very high temperatures and pressures, the core radiates the energy away in the form of neutrinos and within about one second, the neutrinos deposit a fraction of their energy in the infalling material. The outer layers of the star, which are still raining down the material onto the shocked core, feel this increased pressure and reverse the implosion into a supernova explosion (Chevalier 1997). The collapsed core of the massive star forms a neutron star or a black hole, depending on the underlying mechanisms and exact conditions of formation. Analogous to the white dwarf supported by the electron degeneracy pressure, the neutron star equilibrium is maintained by the degeneracy pressure supplied by neutrons and also by the nuclear forces. There is again a limiting mass for a stable neutron star somewhere close to 2-3 M☉ (Baym and Pethick 1979).

The supernova explosions can be broadly classified into two categories: (i) core-collapse, which leaves a remnant behind, (ii) thermonuclear detonation, which entirely disrupts the progenitor star. The first category of supernova explosions can be identified with two types of observational classifications based on their spectra: Type II and Type Ib, Ic. The second category of supernova explosions resulting from an explosive nuclear burning of carbon and oxygen in a white dwarf, are labelled Type Ia by observers. The light curve of Type Ia supernova (Sn) is characterized by a rapid rise to peak brightness within a few days, followed by a steep falloff for about a month, and then a slow decline over several months. Their spectra are characterized by absence of hydrogen lines, but have typically O, Mg, Si, Fe lines. Type Ia Sn are located in both elliptical and spiral types of galaxies. The nature of their progenitors and underlying hydrodynamical systems, however, are still uncertain. The favoured scenario involves a white dwarf near the Chandrasekhar mass limit as a member of a binary with a mass-losing companion. The catastrophic ignition of nuclear fuel induces a thermonuclear explosion of the white dwarf, and a powerful source of energy in the form of radioactive decays most notably of ^{56}Co which is also unstable with a half-life of 77 days to decay into the stable nucleus of ^{56}Fe. These nuclei decay with a release of energy mostly in the high-energy gamma rays, which are scattered by electrons in the envelope of lower energies. The energy so deposited heats up the supernova ejecta and powers the observed light curve of Type Ia, showing an exponential decline (Nomoto, Iwamoto, and Krishimoto 1979).

Type Ia supernova are supposed to be very good, though not accurate, standard candles, since the intrinsic scatter in their peak brightness is within 10–20 per cent. They have, therefore, been profitably employed for inferring the cosmological parameters, although the accuracy of their determination using SNIa would depend on whether the dispersion in the peak brightness is acceptably small and whether a precise calibration of the absolute magnitude of Type Ia supernova can be reliably accomplished. It may very well be that older stellar systems are not made quite the same way as the nearby recent ones and in the younger universe the white dwarfs could have formed with a different chemical composition than now. The supernova light curve would almost certainly be influenced by the amount of gas and dust present in the host galaxy leading to the consequent dimming of its intrinsic brightness. It is also conceivable that SNIa and their galactic environment might evolve with redshift, and nearby supernova light curves might exhibit intrinsic differences from the distant ones. In fact, it has been recently reported (Riess et al. 1999) from comparison of light curves of a dozen nearby Type Ia supernova with those of some 30 high-z Type Ia SN, that the distant supernova appeared to light up more rapidly reaching their peak brightness about two days earlier.

The evolutionary computations indicate that stars with initial masses between 8M$\odot$ and 25M$\odot$ undergo little fallback and are liable to produce a neutron star remnant. The more massive stars in the range 25–40M$\odot$ are, indeed, found to explode, but the amount of mass falling back onto the remnant is sufficiently large to overwhelm the internal forces and induce gravitational collapse to form a black hole. The formation of a neutron star, thus, depends on the requirement that the progenitor massive star is able to eject adequate amount of material without getting completely disrupted and that the remnant is in the permissible range of masses for stable neutron stars. Is there any way to unveil the presence of a black hole resulting from the aftermath of a supernova explosion? The nature of the tail of the light curve generated by the supernova during the late stages, whether it is an exponential or a power-law decline, will reveal the signature of the remnant compact object. A light curve powered by radiative decay, for example, displays an exponential decline in the tail. The accretion onto a black hole has a power-law decay of luminosity with time owing to the existence of an even horizon. In the case of a neutron star, on the other hand, the gravitational energy released during the accretion process is radiated away because of the impact of infalling matter on the hard neutron star surface.

It is evident that condensed objects have been observed with their masses ranging from stellar mass objects formed at the end-point of thermonuclear evolution of massive stars to some others millions or billions of times more massive, which have been detected in the nuclei of galaxies. There is one caveat though, which must be kept in mind: the stellar evolutionary calculations are all based on the implicit assumption that there is only one predominantly controlling parameter, namely, the mass of the progenitor star that is the main determinant of the nature of the remnant. It has been argued (Ergma and van den Heuvel 1998) that the final product of a core-collapse supernova may also depend not only on the progenitor mass, but also on a combination of some additional parameters like rotation and magnetic fields associated with massive stars. Such agents will generate low-frequency electromagnetic radiation, which can provide an additional channel for envelope ejection and might also be responsible for causing asymmetric explosions. In conclusion, the calculations for following the late stages of evolution of massive stars are extremely cumbersome and depend crucially on several physical inputs like nuclear physics, neutrino diffusion mechanisms, shock propagation and neutrino-shock interaction, together with the complex hydrodynamical codes. The final stage has certainly not been reached in our understanding of supernova explosion mechanisms! The relativistic would naturally like to have observational probes of the metric in the strong field domain for testing the space-time structure around Kerr black holes.

The increasing role of strong gravitational fields in driving the energetic phenomena in astrophysics was recognized soon after a succession of discoveries was reported during the second half of the twentieth century. These included violent events in the nuclei of active galaxies, quasars, pulsars, cosmic microwave background radiation, X-ray, and Υ-ray sources. It was immediately realized that the collapsed objects like neutron stars and black holes must provide the underlying basis for powering these spectacular events. The most striking feature of these exotic objects is that in their collapsed state they are endowed with more disposable energy than when they began their life as normal stars. Thus, the gravitational energy of a stellar mass object is

$$E^{\mathrm{grav}} \approx -\frac{GM^2}{R} \approx -10^{48} \text{ ergs}$$

But at the end-point of its thermonuclear evolution, the star can undergo collapse and release a considerable amount of gravitational energy. If a star, for example, were to become a neutron star, its size would shrink by a factor of some 100,000. The gravitational energy released would consequently increase by this factor to something approaching -10^{53} ergs, which exceeds the nuclear energy content of the star. There are two important aspects associated with the process of gravitational collapse: an initially rotating star during contraction, spins even faster in order to conserve angular momentum, thus converting part of its gravitational energy into rotational energy; if the star has an incipient magnetic field, its energy also gets amplified at the expense of gravitational energy. It is this combination of rapid rotation and intense magnetic fields, which is the underpinning mechanism basic to most of the energetic astrophysical phenomena mentioned earlier. The collapsed objects, because of their strong gravitational fields also play an important role in liberating gravitational energy in the process of capturing matter from their environment. In this accretion process a proton falling into the potential well of a neutron star can acquire energy of the order of MeV. Thus, the accretion of gas from a companion or from the surrounding medium onto a compact object turns out to be a very efficient process for generating radiation in X-ray band from the energy of the infalling matter. In Table 4.1 are summarized the physical characteristics associated with the Sun, a White Dwarf, a Neutron Star, and a Stellar Mass Black Hole.

The role of collapsed objects in astrophysics began to get focused sharply only after the discovery of radio pulsars (Hewish et al. 1968). Over 1500

Table 4.1

	Sun	White Dwarf	Neutron Star	Solar Mass Black Hole
Radius (m)	700,000	7000	10	3
Mean density (gm/cm^3)	1	$16^6 - 10^7$	$10^{14} - 10^{15}$	10^{16}
Rotation speed (rad/sec)	3×10^{-6}	3×10^{-2}	$2 \times 10^2 - 10^4$	–
Magnetic field (gauss)	1	$10^6 - 10^7$	$10^{12} - 10^{13}$	–
Efficiency of Gravitational energy Release (GM/Rc^2)	10^{-6}	10^{-4}	$10^{-2} - 10^{-1}$	$6 \times 10^{-2} - 4 \times 10^{-1}$

pulsars are known today and all ∼5000 characterized by periodic brief bursts of radio noise which is quite irregular in almost every respect except for the period between pulses which is remarkably constant. The clock mechanism which keeps the pulsars ticking at such an incredibly constant rate seems to have a satisfactory explanation in terms of rotating stars. However, the emission mechanism whether it occurs at the polar cap or at the light cylinder is still an inadequately understood issue. The discovery of binary pulsar (Hulse and Taylor 1975), 1913 + 16, thought to be a pair of collapsed objects, has revealed the existence of intense gravitational fields that are thousand times stronger than in the solar system. The consequent relativistic effects cause a progressive twisting of the elliptical orbit amounting to a huge 4.2 degrees per year, compared with the 43 seconds of arc per century for the precession of the orbit of planet Mercury. The intense gravity of the imploded double-star system acts as a powerful gravitational transmitter and the orbital period of the binary pulsar is, indeed, observed to be slowly changing with time; the shift in the period of about one part in 10^6 per year is in accordance with the predicted emission of gravitational radiation by the general theory of relativity.

The discovery of binary X-ray sources (Lewin et al. 1995) emitting intense and rapidly flickering X-rays has highlighted the role of accretion from visible companions onto a compact object like a neutron star or a black hole. From the observational standpoint the sources are variable on sufficiently short time-scales without the variations being smeared out across the size of the compact object. On the theoretical front, the gaseous material must fall into a deep enough gravitational potential wall to be heated to X-ray temperatures and to generate sufficiently large luminosities that are observed. The mass of the compact X-ray source can be inferred from the orbital parameters, and knowledge of the limiting mass of a stable neutron star becomes helpful in delineating the nature of the compact member of the binary system. It is gratifying that the inferred masses for periodic X-ray sources are not larger than two solar masses, while the dozen or so irregularly varying binary X-ray sources have much larger masses. There are likely to be tens of millions of stellar mass black holes in our galaxy. One of the outstanding problems of neutron star physics is a reliable determination of the equation of the state of matter over a vast density range spanning 14 orders of magnitude!

Another uncertainty concerns the evolution of neutron star magnetic fields and their location in the interior (Bhattacharya 1995), whether they permeate the whole body or are confined only to the crustal layers of the star. Should the magnetic fields be inherited from the progenitor star, then they are likely to pervade the interior of the neutron star. On the other hand, if the field is generated after its birth via some battery effect or by a dynamo process, it is probably restricted to the outer crustal layers. Most isolated radio pulsars are observed to have high magnetic fields ($10^{11} - 10^{12}$G) with hardly any ohmic decay over a time-scale of $\sim 10^9$yrs. However, the magnetic field of a neutron star appears to decay significantly if it resides as a member of an interacting binary system. It is suspected that the process of accretion from the binary companion onto the neutron star induces the rapid field decay, but the exact decay mechanism is yet to be well understood.

The gamma ray bursts (GRBs), which were first noticed (Mazets 1981) by military satellites some twenty-five years ago, are probably the most dazzling objects in the sky outshining the rest of the universe for a few seconds. Over 3000 GRBs have been detected today. These are characterized by intense bursts of several hundred keV Υ-rays, lasting for a period ranging from about 30 μsec to hundreds of seconds and occurring at the rate of one to two per day. They exhibit extremely irregular light curves with $\sim$ 10 msec fluctuations. The GRBs appear to be isotropically distributed in the sky and are widely believed to be located at cosmological distances. There are now redshift measurements of some half a dozen GRBs available and there is observational evidence from HST for their location in the central regions of star-forming galaxies at moderately high redshifts. The overall energy requirement then amounts to $\sim 10^{52} - 10^{53}$ergs. Another striking feature of GRBs is the detection of afterglow in the radio, optical and X-ray wavebands in several cases. There is another enigmatic class of transient sources called soft gamma-ray repeaters (SGRs) that emit brief (~ 0.1 sec) intense outbursts of low-energy ($\sim 10 - 100$ keV) gamma rays, accompanied by persistent (quiescent) soft X-ray emission (Duncan and Thompson 1992). The SGRs appear to be associated with young ($<10^4$yr) neutron stars embedded in their supernova remnants.

The nature of the central engine responsible for the GRBs and the driver for the ultra relativistic motion from the engine are still the subjects of active investigations (Piran 1998). There are a number of intriguing questions like the bursts occurring with such intensity mainly in the Υ-ray waveband, the associations between GRBs and other energetic events like supernova, and signatures of magnetic fields in the GRB spectra. A number of theoretical proposals to explain GRBs have been advanced which range from crustal quakes on neutron stars in the Galaxy, merging compact objects in distant galaxies, collapsing massive stars (hypernova), etc. The relativistic fireball model seems to account for the efficient transfer of energy by ultrarelativistic motions originating in the central compact source. The resulting shocks produced by highly irregular and intermittent particle fluxes, with faster-moving shells running into slower ones, and the relativistic ejection slowed down by the interstellar medium generating external shocks, seem to account for the observed GRB features like the production of late afterglows. The SGRs are very likely powered by decaying superstrong ($\sim 10^{14} - 10^{15}$ G) magnetic fields associated with some neutron stars and involve both sudden-fracture of the rigid neutron star crust resulting in seismic activity at its surface, and internal heating due to the ohmic decay leading to persistent soft X-ray emission (Kouveliotoy et al. 1998). The GRBs are very important astrophysically since they may represent a class of objects harbouring non-stationary black holes. It is clear there is persistent observational evidence suggesting the presence of collapsed (dark) objects with associated deep gravitational potential wells and horizons through which matter inexorably disappears. What would be exciting, of course, is to search for a signature of the Kerr metric in the strong gravity limit.

A major issue in high-energy astrophysics is to search for an unambiguous signature for black holes. The existence of stellar mass black holes in binary systems may be inferred from the features associated with compact objects like the short-term intensity variation, high efficiency of gravitational energy release in the process of accretion and the resultant high luminosity, mass of the compact source from orbital parameters, the presence of relativistic jets and the accompanying superluminal motions in certain cases, and transient source displaying quasi-periodic oscillations. There are, indeed, black hole candidates in X-ray binaries like CYG X-1, LMC X-3, GRS 1915 + 105, ... Even though black holes and neutron stars are both endowed with deep gravitational potential wells and have similar observational characteristics caused by accretion of matter from their binary companion, there is one major difference and this concerns the inner boundary condition. For a black hole, the accreted matter disappears through the event horizon, carrying most of the viscously dissipated energy with the accreting gas as stored entropy. The light curve is, therefore, expected to show a steep falloff after reaching the peak emission when matter disappears across the horizon. The burst profile of GRS 1915 + 105, which is a superluminal X-ray transient, is highly suggestive of the presence of a black hole. This source studied by Indian X-ray Astronomy Experiment shows sub-second time variability and quasi-periodic oscillations in the X-ray flux (Paul et al. 1998). Broadly, the X-ray properties indicative of a black hole in a binary include ultra-soft spectrum with a high-energy power law tail, bimodal spectral behaviour with the soft component dominant in the high state and hard component in the low state exhibiting fast variability.

A different class of objects are believed to be lurking in the nuclei of galaxies. These are supermassive black holes formed by a variety of routes. The evidence for the existence of supermassive black holes is, of course, indirect and of two kinds: stars in the innermost regions of the nuclei of some galaxies have anomalously large velocities, suggesting the gravitational pull of a central compact mass, in some other galaxies the mass of the hole can be surmised from the properties of the gas with the reprocessed central continuum radiation emission with spectral features.

Various observational lines of evidence suggest that the AGNs and quasars are powered by the accretion of gas onto supermassive ($10^6 - 10^9$ M☉) black holes at the centres of their host galaxies (Rees 1978). The properties of distant quasars and stellar kinematics of nearby galaxies independently suggest the presence of central supermassive black holes in a number of galaxies. There is compelling evidence from the high efficiency of gravitational energy release through disk accretion onto the central black hole, rapid variability of intensity of some AGNs indicating a compact underlying source and presence of superluminal jets, from the optical and radio observations of gas motions in a number of galaxies such as NGC 1068, M87 and M31. The mapping of gas motion with the 1.3-cm maser emission line of H_2O, observations have revealed a rotating disk in the core of NGC 4258 rotating at the speed of over 1000 km/sec around a compact dark mass of about 36 million solar masses. There are indications of the existence of massive black holes (Kormendy and Richstone 1992). There is increasingly convincing evidence for a 2.6 million solar mass black hole lurking deep within the core of our own Galaxy (Genzel and Townesm 1987). However, the characteristic signature of a black hole manifesting as a clean relativistic effect due to its strong gravitational field has only recently been observed (Tanaka et al. 1995) in the case of Seyfert galaxy, NGC 6-30-15. As the accreted gas spirals close to the event horizon, with its velocity approaching the speed of light, the resultant relativistic effects and the gravitational redshift due to the proximity to the black hole, should manifest as X-ray emission from near the inner edge ($\sim 3 - 10$ Rs) of the accretion disk. In fact, an iron $K\alpha$ emission line at ~ 6.4 keV arising from fluorescence is imprinted on the X-ray continuum of many AGNs. The line is extremely broad and asymmetric with most of the flux strongly redshifted, and the width is very likely produced by transverse Doppler and gravitational redshifts close to the central hole implying velocities of the order of a third of the velocity of light.

General relativity has been evidently vindicated in the weak-field limit by astrophysical observations in the solar system and of the binary pulsar. The X-ray emission from a region very close to the black hole is the most direct probe of the strong-field relativistic region. The appearance of the inner region of the accretion disk around a black hole was calculated by Bardeen and Cunningham (1972) taking into account the full general relativistic effects and also the Doppler and Gravitational shifts. The ASCA X-ray satellite was able to measure the emission line profiles in AGNs to demonstrate the broad asymmetric in an emission line indicating the presence of a relativistic disc.

There is mounting evidence for accretion-powered outflows forming jets on the scales ranging from parsecs to hundreds of kiloparsecs. It would seem the jets can be as easily produced by young protostars as by AGNs and are basically formed by accretion discs around central objects, which are threaded by a magnetic field. Thus, the magnetic field as well as the spin of the central object are the underlying features driving the well-collimated jets and in the process removing the excess angular momentum. Following the work of Penrose (1969) who had shown how energy could be extracted from a spinning black hole, Blandford and Znajek (1977) had demonstrated an astrophysical process in which a magnetic field threading a hole could extract the spin energy of the black hole which may be converted into directed Poynting flux and electron-positron pairs. Finally, the orientation of the spin axis of the black hole and the flows in the inner regions of the disk could influence the alignment of the jets and the Lense-Thirring precession would impose the axisymmetry close to the hole (Pringle and Natarajan 1998). The forced precession effects resulting from the torques on the disc can cause the rotation axis to swing over very short time-scales.

Clearly, in future such unambiguous evidence from a number of compact objects in the nuclei of galaxies will establish the existence of putative black holes in the universe at large. The observational astronomy is evidently witnessing a technological renaissance, which is guiding the course of research at the frontiers of astrophysics. The universe is being simultaneously surveyed through a variety of windows from radio to gamma rays. It will be a challenge to speculate if the advances made during this past century will be matched or even surpassed by more exciting technology-driven developments in the ensuing decades.

Rotating supermassive black holes (Kerr black hole)—Signatures

Perhaps, I may digress here to indicate how one may eventually have a confirmation of the space-time around a rotating black hole. If one imagines a Kerr black hole with an accretion disk of free electrons in the equatorial plane, then the polarization of the light emerging from it, after traversing the strong gravitational field of the black hole, will manifest so nonuniform a distribution that should be able to map it. Will Nature be generous enough to provide a clean example which will enable such a mapping?

The X-ray emission, very close to the Kerr black hole, will, probably, be the most direct probe of the relativistic region. Barden and Cunningham have calculated how the inner regions of the accretion disk around the black hole would look like, taking into account the full effects of general relativity. The possible X-ray spectroscopic study of the accretion flow should reveal a signature from the scattered fluorescent line of iron (6.4 keV) and its line profile. The K-alpha line is observed to show a narrow blue wing, Doppler boosted in intensity and a broad wing on the redward side shaped by a combination of transverse Doppler shift and gravitational redshift.

REFERENCES

Alpher, R.N., H.A. Bethe, and G. Gamow (1948). *Phys. Rev.*, 73, 803.

Bardeen, J. and J. Cunnigham (1972). *Astrophys. J.*, 173, L137.
Baym, G. and C.J. Pethick (1979). *Ann. Rev. Astron. Astrophys*, 17, 416.
Bhattacharya, D. (1995). *J. Astrophys. Astro.* 16, 227.
Blandford, R.D, R.L. Znajek, and M.N. Rao (1977). *Royal Astronomical Society*, 179, 433.
Burbidge, E.M., G. Burbidge, W.A. Fowler, and F. Hoyle (1957). *Rev. Mod. Phys.*, 29, 547.
Chandrasekhar, S. (1935). *MNRAS*, 95, 207.
Chevalier, R.A. (1997). *Science*, 276, 1350.
Duncan, R.C. and C. Thompson (1992). *Astrophys. J.*, 392, L9.
Ergma, Ed. and E.P.J. van den Heuvel (1998). *Astron, Astrophys.*, 331, L29.
Genzel, R. and C.H. Townesm (1987). *Ann. Rev. Astron. Astrophys.*, 25, 377.
Hewish, A., S.J. Bell, D.H. Pilkington, P.F. Scott, and R.A. Collins (1968). *Nature*, 217, 709.
Hubble, E. (1934). *Astrophys, J.*, 79, 8.
Hulse, R.A. and J.H. Taylor (1975). *Astrophys. J.*, 195, L51.
Kormendy, J. and D. Richstone (1992). *Astrophys. J.*, 393, 559.
Kouveliotoy, C. et al. (1998). *Nature*, 393, 235.
Kulkarni, S.R. (1997). *Science* 276, 1350.
Lewin, W.H.G., J. van Paradijs, and E.P.J. van den Heuvel (eds) (1995). *X-ray, Binaries*. Cambridge: Cambridge University Press.
Mazets, E.P. et al. (1981). *Nature*, 290, 378.
Nomoto, K., K. Iwamoto, and N. Krishimoto (1997). *Science* 276, 415.
Paul, B. et al. (1998). *Astrophys. J.*, 492, L63.
Penrose, R. (1969). *Rivista del Nuoro Cimento, Numero Speciale* 1, 252.
Piran, T. (1998). *Physics Rep.*, 314, 235.
Pringle, J.E. and P. Natarajan (1998). *Astrophys. J.* 506, 97.
Rees, M.J. (1978). *Observatory*, 98, 210.
Riess, A.G., A.V. Fillipenko, W.L. Schmidt, and B.P. Schmidt (1999). *Astron. J.*, 118, 7668.
Tanaka, Y. et al. (1995). *Nature*, 375, 659.
Weinberger, R. and F. Kerber (1997). *Science*, 276, 1382.

5

THANU PADMANABHAN

Cosmology and Dark Energy

The role of Einstein in shaping our understanding of astronomy and cosmology is tremendous. You only need to remember that if Einstein was wrong in his $E = mc^2$, the sun will not shine! The stars produce their light through nuclear reactions, that is, by exploding large number of hydrogen bombs every second, which, in turn, rely on conversion of mass into energy as taught to us by Einstein. From the stars all the way to the dynamics of the universe, one can see the impact of Einstein at every scale and one could probably give a series of lectures covering it all! But to maintain focus, I will discuss one puzzling aspect of modern cosmology which many people consider to be the *greatest crisis ever faced by theoretical physics*. This problem goes under the name of dark energy and to understand its significance we need to cover a fair bit of ground in cosmology. Accordingly, I will begin with a description of our current understanding of the cosmos and will then describe the issues related to dark energy.

STRUCTURE AND EVOLUTION OF THE UNIVERSE

A look at the night sky might suggest to you that the universe is made of countless billions of stars. But in reality, the stars cluster together as galaxies with a typical galaxy having about a hundred to thousand billion stars. All the stars, which you see in the night sky, belong to our own galaxy called the Milky Way. There are billions of such galaxies strewn across the universe and the nearest large galaxy to Milky Way, called Andromeda, is about 2×10^{24} cm away. Astronomers use a unit called Megaparsec (Mpc) to talk about such large distances with 1 Mpc being equal to roughly 3×10^{24} cm $\approx 3 \times 10^6$ light years. By this unit, the Andromeda galaxy is at about 0.7 Mpc. The farthest galaxies detected by our telescopes are about 5000 Mpc away from us and the mean distance between the galaxies is about 1 Mpc. Cosmologists are usually interested in structures of the scale of a galaxy and bigger and you may say that galaxies are to a cosmologist what atoms are to a solid-state physicist.

Interestingly, the galaxies are not distributed randomly in the sky but are clustered in a well-defined manner. The Andromeda and Milky Way are the largest galaxies in a collection of about 30 galaxies called, rather unimaginatively, the Local Group. A cluster of galaxies could contain anything from 30 to 1000 galaxies, all gravitationally bound together. For example, one of the largest clusters of galaxies called Coma Cluster, which is 300 million light years away, contains over 1000 galaxies. The distribution of galaxies in the sky shows a web-like structure with large voids surrounded by sheets and filaments made of galaxies.

As I mentioned earlier that galaxies are to a cosmologist what atoms are to a solid-state physicist. When you look at the surface of a solid you usually find it to be pretty smooth with no sign of underlying granularity due to individual atoms. This is, of course, because one perceives the solid at length scales much bigger than the separation between two atoms. This idea translates directly to the study of our universe. At the largest scales, the universe appears to be smooth just as at scales much bigger than inter-atomic spacing a solid appears to be smooth. Given the fact that the typical distance between the galaxies is 1 Mpc, we only need to consider regions of the universe of size, say, 200 Mpc for the smoothness to manifest itself. A cubical box of size 200 Mpc kept anywhere in the universe will contain a statistically similar sample of galaxies.

Given this large-scale smoothness, it is convenient to approach the subject of cosmology at two different levels. To begin with, one can consider the very large-scale features of the universe by treating it to be filled with a fairly smooth fluid with specific properties. Having determined the large-scale dynamics, one would like to come back to the formation and evolution of the individual structures like the galaxies. As you will see, this leads to a fascinating picture, which is easy to grasp.

Let us begin with how the universe behaves as a dynamical system filled with matter having a uniform density. Obviously, this description is applicable only at scales, say, larger than 200 Mpc at which one can indeed ignore the granularity in the matter. Once again, in our analogy, this will correspond to studying the properties of a smooth solid with constant density. But unlike in the case of solids, the study of the smooth universe leads to a fascinating discovery. *The universe at the largest scales is expanding!*, that is, any two galaxies in the universe separated by a sufficiently large distance will be moving away from each other with a speed, which increases in proportion to the distance. Typically, two galaxies separated by, say, 500 Mpc will be flying apart with a speed of about 50,000 km/s. (This result, usually attributed to the cosmologist Edwin Hubble, can be deduced from direct observations of distant galaxies. The motion of the galaxy from us leads to a shift in the frequency of the light emitted by the galaxy. This effect, called the Doppler effect, is familiar to all of us—it is the same effect that changes the pitch of an oncoming train with respect to a stationary one.) I would consider the detection of the expansion of the universe as probably *the* discovery of the twentieth century.

The idea that our universe is 'expanding' raises several questions in one's mind and I would like to briefly address two of them: The first question is that if all the galaxies are receding from us, are we at the centre of the universe? The answer is 'no'. The fact that all the distant galaxies are moving away from us does not necessarily mean that we are at the centre of the universe. In fact, if we live on another distant galaxy X, we will again see the same phenomena, viz. that all the galaxies are again moving away from X. To understand this, it is helpful to consider a simpler example. Imagine a very long stretch of road with cars moving in one direction. Assume that, at some instant of time, the distance between any two consecutive cars is 100 metres and that you are sitting in some chosen car, say, A. Let the first car (B) in front of you move at 10 km/hr^{-1} with respect to you, the second one in front (C) at 20 km/hr^{-1}, third (D) at 30 km/hr^{-1}, etc. It will appear to you as though all the cars are moving away from you with the speeds proportional to the distance from you. But consider a man sitting in the car (B) just in front of you. How will he perceive the car just in front of him? Since he is moving at 10 km/hr^{-1} with respect to you and the second car (C) in front of you is moving at 20 km/hr^{-1} with respect to you, the driver of the car (B) will think that the car (C) is moving only at 10 km/hr^{-1} with respect to him. By the same reasoning he will conclude that car (D) in front of him is moving at 20 km/hr^{-1}, etc. In other words, he will come to the *same* conclusion regarding the motion of vehicles as you did! Clearly, there is nothing special about your position. All the cars in this chain enjoy equal status. This is exactly what happens in an expanding universe. Every observer, irrespective of which galaxy he is located in, will see all the galaxies moving away from him with a speed progressively increasing with distance. This curious motion of the galaxies is what really constitutes the 'expansion of the universe'. After all, the word 'universe' is merely a convenient terminology for describing the matter, which we see around us. It is the movement of the parts of the universe which is indicated when one talks about the expansion of the universe.

Since no galaxy in the universe is special, there is no 'centre' of the universe. In the above example, the road is assumed to be infinitely long and any car can be thought of as being in the centre, which is same as saying no car is special and that there is no real centre. The same idea applies in two dimensions or in three dimensions. Think of a rubber sheet with a vertical and horizontal grid of lines drawn on its surface. If the rubber sheet is stretched in the vertical and horizontal directions, the grid will stretch with the rubber sheet. If the sheet is infinite in both the directions, then again there is no point on the rubber sheet, which can be thought of as the 'centre'. All the points are equivalent.

Several popular articles talk about the expanding universe using the analogy of a balloon. As air is blown into the balloon and it expands, any point A on its surface can be thought of as a centre with all other points moving away from the point A. This nicely illustrates the explanation in the previous question, but raises another difficulty: The balloon is expanding into the three-dimensional space. What is the universe expanding into? Is it expanding into higher dimensional space?

The balloon analogy is quite nice but, like all analogies, should be used only for the purpose it was intended. Suppose there are two-dimensional creatures living on the surface of the balloon who are incapable of perceiving the third dimension. If these creatures measure the distance between any two points A and B on the surface of the balloon at two different times, they will find that this distance has increased. The cosmologists among them will interpret it as expansion without ever having to use the third dimension. Similarly, the manner in which we have described the expansion of the universe makes the question 'What is the universe expanding into?' irrelevant. All we know from the observations is that distant galaxies are moving away from us with speeds increasing in direct proportion with the distance. Surely, in an infinite universe such a motion can take place without leading to any contradiction. If we understand the term 'expansion of the universe' as representing 'motion of distant galaxies away from us' then such confusions will not arise.

In the balloon analogy we are assuming that the size of the creatures or their measuring instruments do not expand when the balloon expands, thereby allowing them to make the measurements. In the real universe this question translates into a valid doubt: 'If everything in the universe is expanding (galaxies, stars, earth, metre scales, etc.) then how will we ever know that there is expansion?' This is simply incorrect. Everything in the universe is *not* expanding. All that is happening is an increase in the distances between galaxies, which are sufficiently far away from each other. The expanding model for the universe arises as a solution to the equations in Einstein's theory for gravity when one assumes that the matter is distributed *uniformly*. This is a good assumption at very large scales in the universe just as thinking of a piece of chalk as a continuous solid with uniform density is a good assumption at scales much bigger than the atomic scales. As we probe the piece of chalk at smaller and smaller scales, we will eventually have to recognize the fact that it is made of atoms which themselves are made of electrons, protons and neutrons, etc. One cannot attribute properties of the chalk (like its elastic strength) to the individual atoms, which make up the chalk. Similarly, the expansion of the universe is a property of the universe in the bulk, when the matter is treated as distributed uniformly and the smaller scale structures like galaxies—which are analogous to atoms in a piece of chalk—are ignored. As one proceeds to smaller scales, the existence of these structures needs to be taken into account. In fact, the closest big galaxy to

Milky Way, the Andromeda, is actually moving towards Milky Way rather than moving away from us! This is because the individual gravitational force between Andromeda and Milky Way dominates over the cosmic expansion. The description at small scales needs to take into account the gravitational effect of individual galaxies and we cannot use the results obtained for a uniform distribution of matter. The stars or the earth or the metre scales, which we use for measurements, are *not* expanding and thus can be used to determine the expansion of the universe.

The fact that our universe is expanding implies that it was smaller, denser, and hotter in the past and it evolves with time. In fact, calculations show that some time in the past—about 15 billion years or so ago—the density of the universe was infinite and its volume zero. Fred Hoyle derisively called this event the big bang but the name stuck! We should not think of the big bang as some kind of a Dial cracker explosion, which took place *at a given point in space*. The entire space collapses to a point at big bang rather than space existing independently of matter and the matter collapsing to a point within that space. This, in turn raises further questions: What happened before the big bang? How does one understand scientifically the epoch of the big bang?

We need to remember that close to the 'big bang', the universe was *infinitesimally* small—very much smaller than the size of an atom or even an electron! In trying to understand such a phenomenon we are extrapolating Einstein's theory of gravity, and the cosmological models based on it, to very *small* distances. However, the physics of the subatomic world needs to be described in the language of quantum mechanics. One of the key ideas in quantum mechanics is called 'uncertainty principle'. This principle forbids us from measuring the position of a particle and its speed simultaneously. In the context of the universe, the same principle will prevent us from determining simultaneously the size of the universe and its rate of expansion. It is, therefore, necessary to use a completely different description for understanding the physics of the universe close to the big bang. Until we produce a quantum theoretical description of cosmology, one cannot understand the very early moments, close to the big bang. Unfortunately, we still do not have a good description of the quantum theory of gravity, which incorporates the principles of Einstein's general theory of relativity and the principles of quantum mechanics. Until this is achieved, which is a frontier research area today, we cannot probe what happened to the universe near the 'big bang'. But we do understand a lot about the evolution of the universe since then.

THERMAL HISTORY OF THE UNIVERSE

When the universe expands, the density of matter (and radiation) will decrease. As we go into the past, on the other hand, the universe will become more and more dense. It turns out that the energy contributed by radiation increases faster than the energy contributed by matter as we go into the past. So, at some sufficiently early phase, radiation will dominate over matter.

What was the material content of the universe when it was, say, one-second-old? The temperature of the universe at this time was so high that neither atoms nor nuclei could have existed at this time. Matter must have existed in the form of elementary particles. The exact composition of the universe at that time can be determined by tracking back the contents from the present day. As the universe expands, it cools; when the universe cools to about 1000 million Kelvin, we will be left with a universe containing protons, neutrons, electrons, and photons. Among these particles, protons are positively charged and electrons are negatively charged. They now exist in equal numbers maintaining the overall charge neutrality of matter.

At this epoch, these particles exist as individual entities and not in the form of matter, which we are now familiar with. To create the familiar form of matter, it is first necessary to bring protons and neutrons together and make the nuclei of different elements. Then we need to combine electrons with these nuclei and form neutral atoms. But the typical binding energy of atomic systems is in the range of tens of electron volts. At the temperature of few MeV such *neutral atoms* cannot exist. The binding energy of different atomic *nuclei*, however, is in the range of a few MeV. This suggests that it may be possible to bring neutrons and protons together and form the nuclei of different elements. However, this process—called 'nucleosynthesis'—is also not easy. The key difficulty, of course, is the fact that the universe is expanding very rapidly at this epoch. (When the universe is one-second-old, it will double its size every second; at present the universe is nearly 15 billion years old and hence it will double its size only in another 15 billion years. This shows why the expansion of the universe plays a major dynamical role in the early universe but not at present.) In order to bring particles together in one place we have to overcome the effects of expansion at least temporarily. This is quite impossible to achieve if we try to bring together several nucleons at once. For example, it is not easy to bring two protons and two neutrons together in one place to form the nucleus of helium, which is the second key element in the periodic table. Fortunately, there exists a nucleus called deuterium made of one proton and one neutron with a binding energy of about 2.2 MeV. As the temperature falls below this value, we can combine protons and neutrons and form deuterium nuclei. And then, by combining two deuterium nuclei together, one can form the nucleus of helium. This suggests that the first heavy elements to form out of the primordial soup are deuterium and helium.

In principle, one could try to produce heavier elements by further nuclear reactions. In practice, however, it is not easy to synthesize these atomic nuclei in large quantities, mainly because the expansion of the universe prevents sufficient number of protons and neutrons from getting together. (Stars managed to synthesize heavier elements because they could do it over millions of years at sufficiently high temperatures. In a rapidly cooling universe this is not possible.) Detailed calculations show that only helium is produced in significant quantities. There will be very small traces of deuterium and other heavier elements. The conclusion described above is a key prediction of the standard cosmological model. According to this model, only hydrogen and helium were produced in the early universe. All other heavier elements, which we are familiar with, must have been synthesized elsewhere. We now know that this synthesis took place in the cores of stars, which have sufficiently high temperatures.

As time goes on, the universe continues to expand and cool. Curiously enough, nothing of great importance occurs until the universe is about 400,000 years old, by which time its temperature would have been about 3000 K. At this temperature, matter makes a transition from the plasma state to the ordinary gaseous state with the electrons and ions coming together to form normal atoms of hydrogen and helium. Once these atomic systems have formed, the photons stop interacting with matter and flow freely through space. When the universe expands by another factor of thousand, the temperature of these photons would have dropped to about 3 K or so. Hence we should see all around us today a radiation field with a temperature of around 3 K. The universe today indeed contains such a thermal radiation at a temperature of about 2.7 K! In 1965, Arno Penzias and Robert Wilson at Bell Labs stumbled on the discovery of this radiation. A radio receiver they were developing detected microwave radiation, which did not originate from any of the known sources in the sky. Further studies showed that this radiation was filling the entire universe and was reaching us—so to speak—from the depths of space. It was uniform all over the sky and had a characteristic temperature of about 2.7 K; it was as though the entire universe was enclosed in a box kept at this temperature. This thermal radiation covering the universe is a telltale relic of an earlier hotter phase. As we shall see later, observations related to this radiation play a vital role in discriminating between cosmological models.

The above description of the universe contains the best-understood features of conventional cosmology. You can start the discussion with a universe, which was one-second-old and carry it forward to about 400,000 years. Though this is an impressive feat, it leaves one question unanswered: What will happen to the universe in the future? Will it go on expanding forever? Or will it reach a maximum size and start contracting afterwards?

This situation is analogous to the behaviour of a stone thrown vertically up from the earth. Usually, the stone will rise up to a maximum height and fall back on earth. If we increase the initial speed of the stone, it will reach a higher altitude before falling back. This, however, is true only if the initial speed is below a critical speed called the 'escape speed'. If the stone is thrown with a speed larger than the escape speed, it will escape from the gravity of the earth and will never fall back. In other words, if the gravity is strong enough, it can reverse the speed of the stone and bring it back; but if the gravity is not strong enough, the stone will keep moving farther and farther away from earth. We can compare the expanding universe to a stone, which is thrown from the earth. If the gravitational force of the matter in the universe is large enough, the expansion can be reversed and the universe will eventually start contracting. Since the gravitational force increases with the amount of matter, the fate of the universe depends on the amount of matter present in the universe today. If the matter density is higher than a critical value, this gravitational attraction will eventually win over the present expansion and the universe will start contracting. The critical value of density, which is needed for this re-contraction to take place, can be estimated if we know the speed with which the universe is expanding. It turns out that for our universe to contract back eventually, it should have a matter density of about 5×10^{30} gm cm^{-3} or more. This density is called 'critical density'.

So far so good. Now we need to know the actual mass density of the universe. Is it higher than the critical density or lower? The future of the universe depends on the answer to this question. A partial answer to this question is easy to obtain. We can make a fairly good estimate of the amount of matter which exists in the galaxies, clusters, etc., as long as the matter is emitting electromagnetic radiation of some form: visible light, X-rays or even radio waves. Telescopic observations covering the entire span of the electromagnetic spectrum, along with some theoretical modelling, allow us to determine the amount of visible matter, which is present in our universe. The density contributed by this *visible* matter turns out to be less than about one-tenth the critical density needed to make the universe re-contract. So, if all matter in the universe is visible, then the universe will expand forever.

Unfortunately, this is not the whole story. It turns out that all the matter, which exists in the universe, is *not* visible! In fact, it could very well be that up to 95 per cent of the matter in the universe is 'dark' and does not emit any form of electromagnetic radiation. If this were the case then it is important to identify and estimate the amount of dark matter, which is present in the universe since it is the dark matter, which is governing the dynamics of our universe.

How do we determine the amount of dark matter present in the universe? Astronomers use different techniques for different objects, but the basic principle is to look for gravitational effects, which are *unaccounted* for by the visible matter. Consider, for example, the spiral galaxies. On the outskirts of such galaxies there exist tenuous hydrogen clouds orbiting the galaxy in the plane of the disc. If we see such a galaxy edge-on, then the hydrogen clouds at the two sides of the galaxies will be moving in opposite directions—one towards us and one away from us. These clouds will be emitting radiation, which—because of the Doppler effect—will be redshifted at one end and will be blueshifted at the other. By measuring these shifts, we can determine the speed of these clouds. Knowing the speed and the location of the clouds, one can estimate the amount of gravitating matter present in the galaxy influencing the motion of the cloud.

Astronomers were led to a very surprising conclusion when they performed the above analysis. It turns out that galaxies like ours contain ten times more mass than as determined from the stars and gas. That is, there exists 'dark matter', which is ten times more massive than the visible matter in the Milky Way! Studies carried out for different kinds of galaxies have repeatedly confirmed this result: All galaxies have large dark matter halos around them with the visible part, made of stars etc., forming only a tiny fraction. Most of the gravitational force in the galaxy is due to the dark matter and hence the dynamics of the galaxy is mostly decided by the dark matter halo rather than by the visible matter. There is another surprising thing about the dark matter: it seems to be distributed over sizes nearly three to four times bigger than the visible matter. In fact, astronomers have not been able to determine the 'edge' of the dark matter halo around any of the spiral galaxies. The overall impression one gets from all these observations is that the visible matter is only a tiny speck embedded in the middle of a vast dark matter structure.

What is this dark matter made of? It is possible to put strict bounds on the number density of nucleons (which make up the ordinary matter), which can exist, in our universe. This bound implies that nucleons could at best only account for few per cent of the critical density. The amount of dark matter seen in various systems is far too large to be made of ordinary matter. Most cosmologists believe that the dark matter is made of more exotic particles— particles that interact only very weakly and have been produced in copious quantities in the very early phases of the universe. Several particle physics models postulate the existence of more exotic particles called 'WIMPS' (Weakly Interacting Massive Particles), which could be far heavier than protons, but much less abundant and can be a good candidate for dark matter. There are currently several experiments in progress to detect such exotic particles, though none of them have succeeded yet. A laboratory detection of a dark matter candidate particle will revolutionize our understanding of the universe and establish a remarkable connection between the physics at the smallest and largest scales.

STRUCTURE FORMATION

We saw earlier that our universe contains a hierarchy of structures from planetary systems to clusters of galaxies. Between these two extremes we have stars, galaxies, and groups and clusters of galaxies. How did these structures come into being? After all, one can think of a universe containing a completely homogeneous mass of gaseous hydrogen and helium, expanding steadily forever. In order to form the structures, it is important for the gas to become more concentrated and denser in some regions compared to some other regions. We need a mechanism, which will convert a completely uniform distribution of gas into a distribution punctuated by clumps of matter.

This mechanism happens to be our old friend, gravitational force. The gravitational force can help the growth of non-uniformity in any material medium. Consider now the gravitational force near two points A and B with the actual density being slightly higher in some places (like A) and slightly lower than average in locations (like B). Since the region around A has slightly more mass than the average, it will exert more than average force on the matter around. This will make matter around A contract thereby increasing the concentration of matter around A. In other words, density at regions like B and C will decrease further while the density at A will go up. Now there is even more matter around A; so the gravitational force near A will be still larger causing still more matter to fall on to it, etc. Clearly, this process will make over-dense regions grow more and more dense just because they exert greater influence on the surroundings. This process goes under the name of 'gravitational instability' and is similar to the social situation in which the rich get richer and the poor become poorer. In describing the process of gravitational instability, we have oversimplified matters a little bit. It might appear that the lump at A grows essentially by accreting matter from surroundings. Actually, there is another—more important—process; since the region around A has more than average density, it will also contract under its own weight. That is, the self-gravity of the lump at A will make it collapse further and further which will also increase the density of the lump. (You see that the rich get richer by their own efforts as well as by robbing the poor.) In fact, this process is really the dominant cause for the density to increase at A.

This picture of gravitational instability depends on the existence of small wiggles in the density at some time in the past. If you start with a strictly uniform density, with no lumps like A, then, of course, it will remain strictly uniform in the future. So, we need to have a mechanism, which will produce wiggles in the density distribution in the early universe. In technical language, one says that we must have a mechanism for producing 'density fluctuations'. These wiggles could, of course, be quite tiny; gravity can make them grow eventually, but gravity cannot operate if there were *no* wiggles to begin with.

This description attempts to build the entire universe starting from tiny density fluctuations, which existed, in the earlier phase. You could very well ask whether this model can be verified observationally or whether it is only a theoretical speculation. Fortunately, the gravitational instability model lends itself to a rather stringent test, which it passes in flying colours!

The key to testing the gravitational instability model lies in the microwave background radiation. We saw that when the universe was less than thousand times smaller, matter existed in the form of plasma containing positively charged nuclei of hydrogen and helium and negatively charged electrons. The photons in the universe were strongly coupled to matter and were repeatedly scattered by the charged particles. When the universe cooled further, the electrons and nuclei combined to form neutral atoms and the photons decoupled from the matter. These are the photons, which we see around us in the form of microwave radiation. According to this picture, this radiation is coming to us straight from the epoch of decoupling. It follows that this radiation will contain an imprint of the conditions of the universe in that epoch; by probing this radiation closely, we will be able to uncover the past.

If structures, which we see today, formed out of gravitational instability, then small fluctuations in the density of matter must have existed in the early universe and—in particular—around the time when the universe was thousand times smaller. Since the radiation was strongly coupled to matter around this time, it would have inherited some of the fluctuations in the matter. What does one mean by fluctuations in radiation? Since the thermal radiation is described essentially by one parameter, namely, the temperature, fluctuations in the density would correspond to fluctuations in this temperature. Calculations showed that this temperature difference is extremely tiny, about one part in hundred thousand! But it is vital to look for these temperature fluctuations to verify the theoretical models.

Over the past 14 years several experiments have measured these temperature variations. The first breakthrough came when NASA put into orbit a satellite, called Cosmic Background Explorer (COBE), specifically designed to look for temperature deviations in the microwave radiation. Analysis of data collected by COBE in 1992 has shown that the radiation field does have temperature variations of expected magnitude: about one part in hundred thousand! Several others have now repeated these observations and these fluctuations have been observed with astonishing accuracy. The discovery has caused considerable excitement among cosmologists and—not surprisingly—in the popular press. It shows conclusively that the theoretical models about the formation of the structures in the universe are basically sound. The universe was indeed smoother in the past, but not completely so.

In fact, these measurements have given us a lot more. It tells us a lot about the composition of the universe and—in doing so—leads to the next puzzle.

DARK ENERGY

One key result that emerged from the observations of cosmic microwave background radiation is, the total energy density of our universe is precisely equal to the critical density, but nearly 70 per cent of it must be made of a very smooth dark component, which exerts negative pressure. It is this component, which has been christened dark energy, and for the last several years cosmologists have been trying to come to grips with it.

The two strange features which distinguish *dark energy* from *dark matter* discussed before are the following: First, dark matter behaves like any other material particle, is affected by gravity, and clusters gravitationally. This is the reason why we have halos of dark matter around galaxies and clusters of galaxies exerting gravitational influence on the galaxies. In contrast, dark energy seems to be distributed in a completely smooth manner with almost no wiggles in the density distribution. Second, and more strange, is the fact that dark energy exerts negative pressure on its surroundings. To understand what negative pressure means, let us just contrast it with the normal positive pressure. A balloon filled with normal gas experiences the positive pressure of the gas particles inside. If the balloon is allowed to expand due to the work done by the pressure, its energy will decrease and the gas will cool. But if the balloon was filled with a gas exerting negative pressure, then the expansion need not necessarily decrease its energy content. As the region expands, the energy content can remain constant or even increase if the pressure is sufficiently negative. This is indeed esoteric, but believe it or not, observations show that 70 per cent of the matter in the universe has such an esoteric behaviour.

How could one test this idea directly? Fortunately, there is a way. A universe filled with normal matter will decelerate as it expands. If you compare the expanding universe with a moving car, you can think of three different scenarios for its motion. The car can move along in a particular direction with a steady speed, it can move with an ever-increasing speed (acceleration), or it can move forward with decreasing speed (deceleration). Normal matter in the universe, which has no negative pressure, leads to an expansion with deceleration. On the other hand, the characteristic of matter with negative pressure is that it will lead to an accelerating universe in which the speed of the expansion will keep increasing with time. So, all we need to determine is whether the expansion rate of the universe has increased over time in order to determine whether the universe is dominated by dark energy with negative pressure. Fortunately, this is possible because cosmologists are a blessed lot who can see into the past. Since light takes finite time to go from one point to another, you never observe any object as it is at that instant. For example, when you look at the sun, you see it as it was about eight minutes ago since it takes that much of time for the light to come to you from the sun. Similarly, if you observe a galaxy which is 10 billion light years away, you are seeing it as it was 10 billion years ago! So by observing a sequence of objects at ever increasing distance, one can figure out how the universe was behaving at earlier and earlier times. In particular, one can settle purely observationally whether the universe is accelerating or not.

This was attempted using certain powerful beacons of light called supernova. These are produced when, at the end-stages of their lifecycle, some stars die explosively and brighten up enormously over a period of time. Such supernova, because of their brightness, can be detected up to a very large distance and can be used to determine whether the universe was accelerating or not. These results, obtained over the last several years, suggest that the universe is accelerating. In fact, the observations show that the universe indeed has about 70 per cent of the matter exerting negative pressure exactly as suggested by the microwave background radiation observations.

The fact that two completely different observations have led to the same model of the cosmos has given enormous credibility to these ideas which otherwise would have been treated with much more scepticism.

So, what exactly is this dark energy? The simplest choice for dark energy is something suggested by Einstein himself years back in a completely different context. Einstein found that the equations describing the cosmological model in his theory did not allow a static universe, which lasted from everlasting to everlasting, but instead predicted an expanding universe. Incredibly, Einstein himself hesitated at this point and tinkered with his equations by adding an extra term (called the cosmological constant) in order to obtain a static universe. In doing so, he made two mistakes, first, the term he added did not quite result in a stable static universe and very soon others, notably de Sitter, pointed out that even with the cosmological constant, the universe would continue to expand. The second and more profound error Einstein committed was missing a glorious opportunity to predict the expansion of the universe. In fact, it is this aspect of the cosmological constant, which made him comment later that it was his biggest blunder. It appears now that the cosmological constant really will have the last laugh. If the term, which Einstein had, is present, it will lead to the kind of accelerated universe that we do see.

The difficulty, however, is in explaining or understanding the actual numerical value of this cosmological constant. It turns out that this constant has to be very small and should have a very finely tuned value in order to describe the universe, which we see. A small change in its value will lead to universes, which are totally different from the one we are living in. It has been a challenge to understand why the cosmological constant has the value it has. This perceived difficulty has led researchers to try out other possible candidates for dark energy, each more bizarre than the previous, but none of them has proved to be more successful than the cosmological constant.

6

SANDIP P. TRIVEDI

Einstein's Dream and String Theory

In 1905, Einstein wrote four papers which dramatically altered our understanding of the universe and how it works. Ten years later, in 1915, Einstein made another landmark contribution by formulating his General Theory of Relativity which provided us with a deep understanding of gravity. Einstein was a young man in 1905 only 27 years old. He died in 1955, a full 50 years after, and almost till the end was active as a scientist. In popular science writings Einstein's work till about 1915 is much discussed, but his later work, in the last thirty years of his life, is largely ignored.

In fact, Einstein had a quest that increasingly dominated his later scientific life. His dream was to try and formulate a Unified Theory. A single theory that would bring together all our understanding of nature in one tight mathematical formulation, exquisite in its elegance, and grand in its generality and universality.* The story of how Einstein pursued this dream is a fascinating one, full of courage and tinged with some sadness.

It is a story of courage, because in pursuit of this goal Einstein grew increasingly isolated from the rest of the scientific community. Facing such isolation is never easy even for the most gifted. In the early twentieth century Quantum Mechanics had been developed. In fact, one of Einstein's four papers of 1905 played a key role in this development. Armed with quantum mechanics, physicists now found vast areas of nature accessible to them for study in a quantitatively precise fashion. This was an exciting adventure and it naturally occupied some of the best minds of that time. But Einstein never really accepted quantum mechanics and his path increasingly diverged from those of his fellow physicists.

The story of Einstein's search for a unified theory is one of some sadness, because, unfortunately, Einstein did not live to see his goal being fulfilled.

---

*More correctly speaking, Einstein largely ignored what was being learnt about the nuclear forces in his time. He focused on gravity and electromagnetism and attempted to find a unified theory for them.

Yet, it is eventually also a story of vindication. After Einstein's death the importance of his quest was increasingly realized and his search for the fundamental laws of nature became an integral part of modern day physics. In the mid-1980s this search received a big boost when it was found that string theory had many features that make it a very promising framework to fulfil Einstein's dream.

Here we will discuss, in general terms, what Einstein's quest for unification was about, and some on-going attempts in string theory to fulfil it.

THE LAWS OF NATURE

Let us begin by first recalling that in science we try to understand nature in terms of simple rules or laws. These rules are general and typically apply in many different situations.

As an example consider Newton's Law of Gravitation. Any two objects experience a force between them, due to gravity. The force is attractive and pulls the two towards each other. If the mass of the two objects are M_1, M_2 and if they are separated by distance r the strength of the force is given by

$$F = G_N \frac{M_1 M_2}{r^2} \qquad (1)$$

Here G_N is a constant, called Newton's constant, see Fig. 6.1. This is a very general rule which is universally valid. It does not matter what the two objects are made out of. One of them could be the earth and the other an apple. Or one of them could be made of wood and the other of wax. Or one of them could be the earth and the other the sun or the moon. In all these different situations the same rule applies. Similarly, it does not matter where the objects are located. They could be on earth, or near earth in the solar system, or far away in the Milky Way.

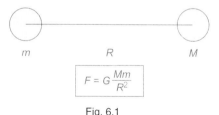

Fig. 6.1

Newton's Law is remarkably simple and elegant. In fact, it is precisely its simplicity which makes the law so universally valid. Had the law been less universal, say with one set of rules for objects made up of wood, another for apples, etc., then the rules would have had to depend on the 'woodness' or 'appleness' of the objects and on all the other attributes which when changed would change the rules. Such a law would have automatically been more complicated and less elegant.

Newton's Law can be used to predict the outcome of several experiments before they are performed. This allows the law to be tested in a number of different situations. The law, we saw above in Eq. (1), depends on one constant, G_N. Once this constant is measured in a few experiments, we can then use the Law to predict what will happen in many different situations.[1] Newton's Law has passed many such experimental tests repeatedly and in many different situations, this gives us considerable confidence that it is correct. Now a little thought shows that it is, in fact, the simplicity of the law which gives it so much predictability. Had the law been more complicated, there would have been additional unknowns, besides G_N, which would have had to be determined before predictions could be made.

To summarize, Newton's Law of Gravitation is an example of one of the important laws of physics. It is simple and elegant, and this simplicity is responsible for its universality and its great predictive power. These features are common to all the important laws in physics.

In the history of physics, it has been found that as our understanding of nature has advanced, the rules or laws which were known to explain different phenomena have come together in the form of even more general laws. These more general laws have provided an underlying explanation for the earlier discovered laws and the different phenomena explained by these laws. In this was many different phenomena in nature have been found to have a common or unified explanation.

An example of this is given by Maxwell's theory of Electromagnetism. This theory provides the underlying explanation for all phenomenon in classical physics occurring due to the forces of electricity and magnetism. Prior to Maxwell's formulation of this theory it was not recognized that the forces of electricity and magnetism had a common origin of this type. Once again, as in the case of Newton's Law of Gravitation, this theory is simple, and universal with great predictive power.

EINSTEIN'S SEARCH FOR A UNIFIED THEORY

It was in this historical context that Einstein began to wonder if a unified theory could be formulated. By this Einstein meant a single set of laws, with no arbitrariness, on the basis of which all the different phenomena in nature could be explained. This single set of laws would thus provide a unified description for all the phenomena in nature.

Here is a quotation from Einstein himself asserting his belief that such a fundamental set of laws does exist.

'Nature is constituted so that it is possible to lay down such strong determined laws that within these laws only rationally, completely determined constants occur, not constants therefore which could be changed without completely destroying the theory.'

The constants Einstein refers to above are the analogue of Newton's gravitational constant, G_N which we saw appears in Newton's Law of

Gavitation. Einstein's goal was lofty. If such constants occurred in the final unified theory, he said they must have an explanation, they cannot be arbitrary.

Einstein spent 30 years searching for the unified theory. As mentioned in the introduction that he was unsuccessful. Here, we will not explore in detail the different turns and twists that his search took. One aspect of his thinking is worth emphasizing though. Einstein never quite accepted quantum mechanics. He searched for a unified theory that was deterministic hoping that quantum mechanics with its probabilistic aspects would emerge from this more fundamental description.

In the years following Einstein's death the importance of his quest for a unified theory was increasingly realized by many physicists. By the late 1970s the standard model of particle physics was well understood. It was known that there are four fundamental forces of nature, gravity, and electromagnetism, which we have already mentioned above, and two other forces, the strong force and the weak force, which are important in nuclear physics. Of these electromagnetism, the strong and weak forces were reasonably well understood and it was known that the underlying laws describing them were based on quantum mechanics. With this basic understanding in place the attention of physicists increasingly turned to Einstein's search for a unified theory.

Since quantum mechanics had been so successful by then, Einstein's attempts to give up on it seemed out of place. It seemed more reasonable to try and formulate a set of laws compatible with quantum mechanics that might provide an underlying explanation for all four fundamental forces of nature. However, attempts in this direction proved unsuccessful.

The central difficulty was in putting gravity together with quantum mechanics. To understand the difficulty let us go back to Newton's Law of Gravitation. This is a classical law which does not include the effects of quantum mechanics. Once quantum effects are included the law would be altered so that Eq. (1) takes the form,

$$F = G_N \frac{M_1 M_2}{r^2} + \text{corrections} \qquad (2)$$

The first term on the right hand side of this equation was present in eq. (1), the second term which are corrections, arise due to quantum effects. Now since Newton's Law is known to work very well, and agrees with many experiments, the corrections in all the situations where the law applies should be small. However, when physicists in their attempt to formulate a unified theory tried putting quantum mechanics together with gravity, they inevitably found that the corrections in eq. (2) were very big. In fact, the corrections turned out to be infinitely large! Large corrections would have been enough to show that such a theory is in contradiction with experiments and hence incorrect. But the fact that the corrections were infinitely large showed that the theory was mathematically inconsistent, which is a complicated way of saying that it did not make sense.

Several attempts at finding a unified theory, over many decades, met with failure as, mentioned above.

STRING THEORY

In the 1980s it was realized that a great deal of progress could be made if one entertained a bold hypothesis. The hypothesis was to assume that everything is made of strings. The theory describing how these strings behaved was worked out to some extent, and is called string theory. It was found that string theory has several attractive features that make it a remarkably promising candidate for realizing Einstein's dream of a unified theory. Naturally these developments got many physicists very excited.

To understand why the string hypothesis is so bold, let us pause to discuss some of the developments in the preceding years better. As mentioned above, by the late 1970s, the workings of the electromagnetic, strong and weak forces were understood to good extent. These developments revealed that matter when examined at small sizes is made up of particles. For example, the different forms of matter we see around us are all made up of atoms and atoms in turn are made up of electrons, protons, and neutrons. Further progress had revealed that neutrons and protons contained even smaller particles called quarks, etc. So from this perspective it was natural to believe that the ultimate, or smallest, constituents of matter would also be particle like. And indeed this was the basis for the preceding attempts at unification.

The underlying hypothesis in string theory is different though, it says that matter is actually made up of tiny strings. Electron, quarks, photons, etc., are all strings. In fact, they are all the same kind of string. This one string can vibrate in many different modes, just as a sitar string can be plucked to give many different notes (see Fig. 6.2). These different notes of vibration correspond to the different kinds of particles, the electron, quarks, photons, etc. The strings are very tiny in size, so they appear like particles, but if we can construct a powerful enough microscope we will see that they are actually strings.

Now it is important to emphasize that this is a hypothesis. To prove that it is true one needs to build a very powerful microscope and examine the constituents of matter with high enough resolution to see if they are indeed made out of string. This is not an easy task and has not been done as yet.[2] However, theoretically this hypothesis can be explored, and what has been found is quite exciting.

One starts with this hypothesis and also assumes that the strings interact with each other in the framework of quantum mechanics. This starting point then automatically gives rise to gravity and also to forces like electromagnetism and the weak and strong interactions. Thus string theory unifies our description of these different forces of nature. The essential reason

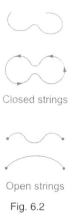

Fig. 6.2

for this unification is that in string theory all the elementary particles are made up of the same fundamental object—a string vibrating in different ways.

The corrections to Newton's Law, which we discussed in eq. (2), come out to be indeed small in string theory. Thus a major stumbling block in all previous attempts is overcome.

String theory has a number of other successes. It has been known since the 1970s, due to the work of Hawking and Beckenstein and many others, that black holes have an entropy, just like a cylinder full of hot gas. For gas it is well-known that this entropy arises due to the jiggling of atoms and molecules. The jiggling makes the molecules move helter-skelter in a disordered manner, and their entropy is a measure of this disorder. But for black holes the source of this disorder was a big puzzle. In string theory this puzzle has been understood to some extent. The entropy for a class of black holes has been calculated from first principles in the theory and agree with the results of Hawking and Beckenstein.

We do not want to give the impression here that everything is well-understood in string theory. Far from it, string theory is a work in progress and there are several open important questions. One of the most important concerns the fact that the theory requires space to have extra dimensions. This seems at first crazy, after all we know that space has three dimensions, how can there be any more? It turns out that there can be extra dimensions if they are small in size. A good analogy is provided by a pencil or a cylinder of small diameter. The surface of the pencil or the cylinder is two dimensional. But when viewed from afar it looks like a line, which is one dimensional, because the extra dimension cannot be seen. Similarly, the extra dimensions of string theory, if small, would have escaped observation as yet. More careful experiments, which probe very short distances, would have to be done for this.

The idea that extra dimensions, which are small, can exist and that this would help with unification is actually a much older one, and due to the imaginative work of Kaluza and Klein in 1920s. In fact, Einstein himself really liked this idea and tried to use it in his attempts at fashioning a unified theory. In string theory this idea finds a natural place. But we still need to understand in string theory what size these extra small dimensions take, what shape they take, and how experimentally may we test if this is indeed what is going on. This is one of the outstanding questions in the subject.

Before we move on, let us discuss that attempts to incorporate Kaluza and Klein's ideas in more conventional approaches had run into an important problem. It is known that the matter in the standard model of particle physics has a handedness or chirality to it. If we consider some nuclear reactions involving the weak interactions for example, and look at their image in a mirror, a physicist will be able to tell the reaction happening in the real world from the fake image. It was shown to be impossible to obtain this handedness in conventional approaches if there were extra dimensions that were curled up along the lines of Kaluza and Klein's ideas. But in string theory this can be overcome and matter quite akin to that in the standard model can be found. This is another example of how String Theory has been able to successfully overcome challenges with more conventional approaches to unification.

RECENT PROGRESS IN COSMOLOGY AND STRING THEORY

Over the past decade there has been some remarkable progress in observational cosmology. This progress has revealed some very interesting new facts and puzzles about the universe. Since these developments merit a whole information in themselves and, in fact, since such an information was given by Professor Padmanabhan in his lecture, we will only briefly review some of these developments here. Our main purpose is to discuss how these developments are in turn connected to on-going developments in string theory. Understanding some of the puzzles the experiments in cosmology have thrown up is forcing us to improve our understanding of string theory and that in turn is feeding back and leading, to some extent, to a better understanding of these puzzles in cosmology.

It has been known since the work of the astronomer Edwin Hubble that the universe is expanding. The rate of expansion is measured by the Hubble constant, H, which is approximately 70 km/sec/Megaparsec. One parsec is about 3×10^{18} cm. A Megaparsec is six orders of magnitude bigger. This means that in the universe two galaxies which are separated by one Megaparsec will grow further a part by about 70 km in one second. If the galaxies are separated by a smaller distance, then they will grow apart by a proportionately smaller amount. On terrestrial scales where distances are tiny and time scales small compared to cosmological scales, this expansion is too small to have any significance. But when dealing with galaxies separated by huge distances the expansion has a significant observable effect and is important.

A good way to visualize this expansion is to picture the universe as the surface of a balloon, see Fig. 6.3. Galaxies are dots on the surface of this balloon. As time goes by the balloon is blowing up, its surface is expanding and the dots which are galaxies are flying apart from each other.

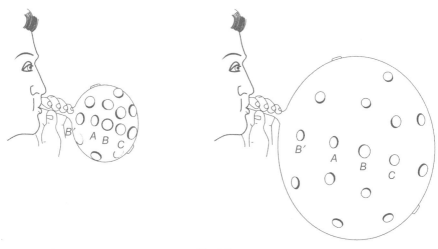

Fig. 6.3

The expansion of the universe is old news. Over the past decade several new experiments have been carried out. Some of these have studied the cosmic microwave background radiation, that comes to us from very early in the history of the universe, very precisely. Other experiments have looked at supernovae, which are very bright sources, that can be observed from very far away. These experiments have revealed something very puzzling: namely that the rate at which the galaxies are flying apart is increasing with time. That is to say, the galaxies are flying apart from each other a little more rapidly today than they were a hundred million years ago.

This discovery is puzzling for the following reason: gravity is known to be universally attractive. Thus it was widely expected that the rate of expansion should be slowing down. Galaxies should exert an attractive gravitational force on each other, this should slow down the rate at which they fly apart from each other not speed this rate up. In terms of the analogy suggested above the speed at which the balloon is blowing up should be decreasing due to gravitational attraction not increasing as the experiments show us.

The speed-up in the expansion rate must be because some mysterious new form of matter-energy is dominating the universe, at large enough distance, causing in effect gravity to repel and leading to acceleration. Nobody knows what this mysterious form of matter-energy really is. We have given it a name, it is called Dark energy. Astronomers estimate that about 70 per cent of the matter-energy in the universe is made up of Dark energy.

Theoretically, the most plausible explanation is something called the cosmological constant. In fact, it was Einstein himself who first entertained the possibility that a cosmological constant might exist, but the data in his time was not very good and seemed to suggest no evidence for the idea. Einstein quickly abandoned it calling this 'his biggest blunder'. However, now that the data is better, we find that this is no blunder. In fact, a cosmological constant of positive sign and appropriate magnitude could explain the mysterious repulsion of gravity at large distance and the resulting acceleration of the expansion rate.

THE COSMOLOGICAL CONSTANT AND STRING THEORY

The experiments discussed in the previous section then give rise to an important question that any fundamental theory must face up to. Can this mysterious repulsion leading to acceleration at cosmological scales arise in a fundamental theory of gravity? If we accept that the repulsion is due to a positive cosmological constant, the question becomes: whether a positive cosmological constant can arise in a consistent fundamental theory of gravity?

In particular this is a question of importance for string theory, which we have argued above is exactly such a promising consistent fundamental theory for gravity and the other forces of nature.

Preliminary attempts, in fact, suggested that it was not possible to obtain a positive cosmological constant in string theory. However, in research with Shamit Kachru, Renata Kallosh, and Andre Linde (2003), and based on earlier work with Prasanta Tripathy, etc., we showed that this is not true. In fact, a positive cosmological constant and thus an accelerating universe can indeed arise in string theory.

The question of whether an accelerating universe can arise was in effect an important test for string theory and our result shows that it passes it well. Just like the other successes of string theory, for example, the fact that it can give rise to forces and matter quite akin to what is seen in nature, here too we see that the theory possesses the required structure to account for an important feature of the universe—its acceleration.

The result that a positive cosmological constant can arise is thus of considerable significance for the study of string theory. It is also of some significance for cosmology. Had a positive cosmological constant not been possible to obtain from a more fundamental starting point which incorporates both gravity and quantum mechanics, one would be tempted to think that there is some explanation for the acceleration, other than the cosmological constant.

There is an important issue we have not addressed so far. We had mentioned that a positive cosmological constant gives rise to an accelerating universe and we have seen that a cosmological constant of positive sign can arise in string theory. But the experiments I mentioned have also measured the magnitude of this acceleration and one should ask whether this measured value is in agreement with the calculations in string theory.

The answer is a bit of a disappointment. One finds that the theory does not predict a unique value for the cosmological constant. In fact, it can be positive with widely varying values, which would correspond to universes that were accelerating at widely varying rates, typically much more quickly than ours; it can also give rise to negative cosmological constants, which would correspond to universes which are not accelerating at all.

There is, in fact, a rich and complicated landscape in string theory, consisting of many different possibilities. We schematically depict this in Fig. 6.4. The minima at the bottom of each trough roughly correspond to possible states the universe can be in today, and the height of the minimum above the X-axis tells us what the cosmological constant is in any particular state. If the minimum lies above the X-axis it has a positive cosmological constant, if it lies below it has a negative cosmological constant. Now, actually this figure is actually a vast simplification. In string theory there are about a 100 directions, instead of just one X-axis, and the total number of ground states is estimated to be mind-blowingly large, about 100^{500} or so—although this estimate is a bit preliminary! Just as the cosmological constant varies in these different minima, so do other constants of nature, like G_N—Newton's gravitational constant that we had discussed in the beginning of this chapter. In fact, the nature of the fundamental forces themselves can vary, besides the constant which tells us about their strength, as one moves in the landscape and goes to different minima. This means the theory allows for many different kinds of universes, some of them are accelerating, others are not, and with different rates, etc.

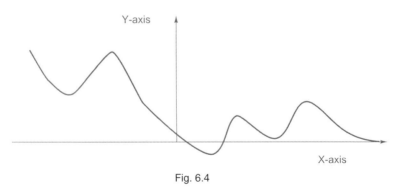

Fig. 6.4

This embarrassment of riches throws up a big challenge. How is string theory going to meet Einstein's challenge of being a unified theory with no arbitrariness in its description of nature? And how are we going to be able to get concrete predictions from the theory if it allows for so many choices and so many possibilities? These questions are matters of intense discussion and debate amongst scientists at the moment. Hopefully further progress will shed more light on these issues.

CONCLUSION

Einstein had a dream, a central quest which shaped his later scientific life. This was to find a unified theory, that would describe all the different phenomena in nature. We have seen in this chapter that the search for such a theory has gained new impetus in the past few decades with the advent of string theory. This theory is based on the assumption that everything in nature is made up of little strings. With this starting point, and based on quantum mechanics, one finds that a universe with gravity and other forces of nature quite akin to ours naturally arises in the theory.

Recent experiments in cosmology have shown that the universe is accelerating. This had thrown up a new challenge for string theory, namely, whether this unexpected acceleration can arise in string theory. Our results have answered this question in the affrmative by showing that a positive cosmological constant can arise in string theory. But these results in turn have raised many questions. They have shown that there is not only one allowed universe in string theory but, in fact, many, with widely varying values for the rate of acceleration and also many other properties of nature. How can string theory be turned into a predictive theory in light of this vast set of possibilities? And how can it hope to fulfil Einstein's dream of a unified theory with no arbitrariness in the choice of the constants of nature? These are questions which we must now grapple with.

It is clear that the importance of Einstein's quest is now well recognized and an important activity in physics. It was not so during Einstein's life. For much of the 30 years that he searched for a unified theory, he worked nearly alone, walking a difficult path almost by himself with courage and conviction. He was an embodiment of the ideals that Tagore expressed in his poem, '*Ekla Cholo Re*', 'Walk alone. Even when there is no one else to walk with you, walk alone'.

The search for a unified theory has entered a truely exciting phase now. Young people, barely having finished their PhDs, are making some of the most notable contributions in this subject. Some of the most original work in the area of string theory is being done in India. Invite young people to learn more about these developments and to join in this search. This is perhaps the best tribute we can pay to Einstein on the hundredth anniversary of his miraculous year.

NOTES

1. Some tests of the law require us to know the masses of the two objects involved and the distance between them. These parameters would have to determined by independent experiments to test the law. There are other consequences of the law, for example, the fact that planets move in elliptical orbits around the sun, which do not depend on knowing all these parameters.
2. The most powerful microscopes are actually giant colliders that smash elementary particles, like electrons, or protons, together at very high energies.

In 2008, the biggest of these accelerators will start functioning, at CERN in Geneva, with Indian participation. Scientists hope an exciting era will unfold, where we will learn a great deal that should help in our search for a unified theory.

REFERENCES

Balaram, P. (ed.) (2005). *Current Science*. Volume 89, Number 12, 25 December.

——— (2004). *Current Science*. Volume 87, Number 12, December.

Greene, Brian (2000). *The Elegant Universe*. New York: Vintage Books.

Kachru, S., R. Kallosh, A. Linde, and S.P. Trivedi (2003). 'De Sitter Vacua in String Theory', *Physical Review D*, Volume 68, 046005.

Pais, Abraham (1983). *Subtle is the Lord: The Science and the Life of Albert Einstein*. New York: Oxford University Press.

7

ABHAY VASANT ASHTEKAR

Space and Time
From Antiquity to Einstein and Beyond

FROM ANTIQUITY TO EINSTEIN

> As an older friend, I must advise you against it, for, in the first place you will not succeed, and even if you succeed, no one will believe you.
>
> Max Planck to Albert Einstein, on learning that Einstein was attempting to find a new theory of gravity to resolve the conflict between special relativity and Newtonian gravity (1913)

Every civilization has been fascinated by notions of Space (the Heavens) and Time (the Beginning, the Change, and the End). Early thinkers from Gautama Buddha and Lao Tsu to Aristotle commented extensively on the subject. Over centuries, the essence of these commentaries crystallized in human consciousness, providing us with mental images that we commonly use. We think of space as a three dimensional continuum which envelops us. We think of time as flowing serenely, all by itself, unaffected by forces in the physical universe. Together, they provide a stage on which the drama of interactions unfolds. The actors are everything else—stars and planets, radiation and matter, you and me.

For over a thousand years, four books which Aristotle wrote on physics provided the foundation for natural sciences in the Western world. While Heraclitus had held that the universe is in perpetual evolution and everything flowed without beginning or end, Parmenides had taught that movement is incompatible with Being which is one, continuous, and eternal. Aristotle incorporated both these ideas in his cosmogonic system. Change was now associated with the earth and the moon because of imperfections. Changelessness was found on other planets, the sun and stars because they are perfect, immutable, and eternal. In modern terms one can say that in

Aristotle's paradigm, there was absolute time, absolute space, and an absolute rest frame, provided by earth. This was the reigning world-view Isaac Newton was exposed to as a student at Cambridge in 1661–5.

Twenty years later, Newton toppled this centuries old dogma. Through his *Principia*, first published in 1686, he provided a new paradigm. Time was sill represented by a one-dimensional continuum and was absolute, the same for all observers. All simultaneous events constituted the three-dimensional spatial continuum. But there was no absolute rest frame. Thanks to the lessons learned from Copernicus, earth was removed from its hitherto privileged status. Galilean relativity was made mathematically precise and all inertial observers were put on the same physical footing. The *Principia* also shattered Aristotelian orthodoxy by abolishing the distinction between heaven and earth. Heavens were no longer immutable. For the first time, there were universal principles. An apple falling on earth and the planets orbiting around the sun were now subject to the same laws. Heavens were no longer so mysterious, no longer beyond the grasp of the human mind. In the beginning of the 1700s, papers began to appear in the Proceedings of the Royal Society, predicting not only the motion of Jupiter, but even of its moons! No wonder then that Newton was regarded with incredulity and awe not only among lay people, but even among leading European intellectuals. For example, Marquis de l'Hôpital—well-known to the students of calculus for the l'Hôpital rule—eagerly wrote from France to John Arbuthnot in England about the *Principia* and Newton: 'Good god! What fund of knowledge there is in that book? Does he eat and drink and sleep? Is he like other men?' As Richard Westfall put it in his authoritative biography of Newton, *Never at Rest*,

Newton was hardly an unknown man in philosophic circles before 1687. Nevertheless, nothing had prepared the world of natural philosophy for the *Principia*. A turning point for Newton, who, after twenty years of abandoned investigations, had finally followed an undertaking to completion, the *Principia* also became a turning point for natural philosophy.

The *Principia* became the new orthodoxy and reigned supreme for over 150 years. The first challenge to the Newtonian world-view came from totally unexpected quarters: advances in the understanding of elecromagnetic phenomena. In the middle of the nineteenth century, a Scottish physicist James Clarke Maxwell achieved an astonishing synthesis of all the accumulated knowledge concerning these phenomena in just four vectorial equations. These equations further provided a specific value of the speed c of light. But no reference frame was specified. An absolute speed blatantly contradicted Galilean relativity, a cornerstone on which the Newtonian model of space-time rested. By then most physicists had developed deep trust in the Newtonian world and, therefore, concluded that Maxwell's equations can only hold in a specific reference frame, called the ether. But, they reverted back to the Aristotelian view that Nature specifies an absolute rest frame. A state of confusion remained for some 50 years.

It was the 26-year-old Albert Einstein who grasped the true implications of this quandary: It was crying out, asking us to abolish Newton's absolute time. Einstein accepted the implications of Maxwell's equations at their face value and used simple thought experiments to argue that, since the speed c of light is a universal constant, the same for all inertial observers, the notion of absolute simultaneity is physically untenable. Spatially separated events which appear as simultaneous to one observer cannot be so for another observer, moving uniformly with respect to the first. The Newtonian model of space-time can only be an approximation that holds when speeds involved are all much smaller than c. A new, better model emerged and with it new kinematics, called special relativity. Time lost its absolute standing. Only the four-dimensional space-time continuum had an absolute meaning. Space-time distances between events are well defined, but time intervals or spatial distances between them depend on the state of motion of the observer, that is, of the choice of a reference frame. The new paradigm came with dramatic predictions that were hard to swallow. Energy and mass lost their identity and could be transformed into one another, subject to the famous formula $E = mc^2$. The energy contained in a gram of matter can therefore illuminate a town for a year. A twin who leaves her sister behind on earth and goes on a trip in a spaceship travelling at a speed near the speed of light for a year would return to find that her sister had aged several decades. So counter-intuitive were these implications that as late as the 1930s philosophers in prominent Western universities were arguing that special relativity could not possibly be viable. But they were all wrong. Nuclear reactors function on earth and stars shine in the heavens, converting mass into energy, obeying $E = mc^2$. In high-energy laboratories, particles are routinely accelerated to near light velocities and are known to live orders of magnitude longer than their twins at rest on earth.

In spite of these revolutions, one aspect of space-time remained Aristotelian: It continued to be a passive arena for all 'happenings', a canvas on which the dynamics of the universe are painted. In the middle of the nineteenth century, however, mathematicians discovered that Euclid's geometry that we all learned in school is only one of many possible geometries. This led to the idea, expounded most eloquently by Bernhard Riemann in 1854, that the geometry of physical space may not obey Euclid's axioms—it may be curved due to the presence of matter in the universe. It may not be passive but could act and be acted upon by matter. It took another sixty-one years for the idea to be realized in detail.

The grand event was Einstein's publication of his general theory of relativity in 1915. In this theory, space and time fuse to form a four-dimensional continuum. The geometry of this continuum is *curved* and the amount of curvature in a region encodes the strength of the gravitational field there. Space-time is not an inert entity. It acts on matter and can be acted upon. As the American physicist John Wheeler puts it: *Matter tells space-time how to bend* and *space-time tells matter how to move*. There are no longer any spectators in the cosmic dance, nor a backdrop on which things happen. The stage itself joins the troupe of actors. This is a profound paradigm shift. Since all physical systems reside in space and time, this shift shook the very foundations of natural philosophy. It has taken decades for physicists to come to grips with the numerous ramifications of this shift and philosophers to come to terms with the new vision of reality that grew out of it (Ashtekar 2005).

GRAVITY IS GEOMETRY

It is as if a wall which separated us from the truth has collapsed. Wider expanses and greater depths are now exposed to the searching eye of knowledge, regions of which we had not even a pre-sentiment.

Hermann Weyl, on 'General Relativity'

Einstein was motivated by two seemingly simple observations. First, as Galileo demonstrated through his famous experiments at the leaning tower of Pisa, the effect of gravity is universal: all bodies fall the same way if the only force on them is gravitational. Second, gravity is *always* attractive. This is in striking contrast with, say, the electric force where unlike charges attract while like charges repel. As a result, while one can easily create regions in which the electric field vanishes, one cannot build gravity shields. Thus, gravity is omnipresent and non-discriminating; it is everywhere and acts on everything the same way. These two facts make gravity unlike any other fundamental force and suggest that gravity is a manifestation of something deeper and universal. Since space-time is also omnipresent and the same for all physical systems, Einstein was led to regard gravity not as a force but a manifestation of space-time geometry. Space-time of general relativity is supple and can be visualized as a rubber sheet, bent by massive bodies. The sun for example, being heavy, bends space-time enormously. Planets like earth move in this curved geometry. In a precise mathematical sense, they follow the simplest trajectories called geodesics—generalizations of straight lines of the flat geometry of Euclid to the curved geometry of Riemann. So, when viewed from the curved space-time perspective, earth takes the straightest possible path. But since space-time itself is curved, the trajectory appears elliptical from the flat space perspective of Euclid and Newton.

The magic of general relativity is that, through elegant mathematics, it transforms these conceptually simple ideas into concrete equations and uses them to make astonishing predictions about the nature of physical reality. It predicts that clocks should tick faster in Kathmandu than in Mumbai. Galactic nuclei should act as giant gravitational lenses and provide spectacular, multiple images of distant quasars. Two neutron stars orbiting around each other must lose energy through ripples in the curvature of space-time caused by their motion and spiral inward in an ever tightening embrace. Over the last thirty years, astute measurements have been performed to test if these and other even more exotic predictions are correct. Each time, general relativity has triumphed (Will 1986). The accuracy of some of these observations exceeds that of the legendary tests of quantum electrodynamics. This combination of conceptual depth, mathematical elegance, and observational successes is unprecedented. This is why general relativity is widely regarded as the most sublime of all scientific creations (Chandrasekhar 1991, ch. 7).

BIG-BANG AND BLACK HOLES

The physicists succeeded magnificently, but in doing so, revealed the limitation of intuition, unaided by mathematics; an understanding of Nature, they discovered, comes hard. The cost of scientific advance is the humbling recognition that reality is not constructed to be easily grasped by the human mind.

Edward O. Wilson, *Consilience: The Unity of Knowledge*

General relativity ushered in the era of modern cosmology. At very large scales, the universe around us appears to be spatially homogeneous and isotropic. This is the grandest realization of the Copernican principle: our universe has no preferred place nor favoured direction. Using Einstein's equations, in 1922 the Russian mathematician Alexander Friedmann showed that such a universe can not be static. It must expand or contract. In 1929, the American astronomer Edwin Hubble found that the universe is indeed expanding. This in turn implies that it must have had a beginning where the density of matter and curvature of space-time were infinite. This is the *big-bang*. Careful observations, particularly over the last decade, have shown that this event must have occurred some fourteen billion years ago. Since then, galaxies are moving apart, the average matter content is becoming dilute. By combining our knowledge of general relativity with laboratory physics, we can make a number of detailed predictions. For instance, we can calculate the relative abundances of light chemical elements whose nuclei were created in the first three minutes after the big-bang; we can predict the existence and properties of a primal glow (the cosmic microwave background) that was emitted when the universe was some 400,000 years old; and we can deduce that the first galaxies formed when the universe was a billion years old. An astonishing range of scales and variety of phenomena!

In addition, general relativity also changed the philosophical paradigm to phrase questions about the Beginning. Before 1915, one could argue—as Immanuel Kant did—that the universe could not have had a finite beginning. For, one could then ask, 'what was there before?' This question pre-supposes that space and time existed forever and the universe refers only to matter. In general relativity, the question is meaningless: since space-time is now *born* with matter at the big-bang, the question 'what was there *before*?' is no longer meaningful. In a precise sense, big-bang is a boundary, a frontier, where

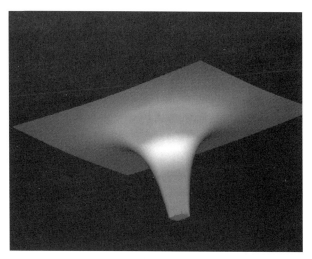

Fig. 7.1 A depiction of the universe originating at the big-bang and then expanding. Time runs vertically. In general relativity the curvature becomes infinite at the big-bang, tearing the very fabric of space-time continuum. The smooth conical surface depicts expanding space-time and the ragged edge at the bottom depicts the tearing of the fabric at the big-bang (courtesy Dr Pablo Laguna).

space-time ends. General relativity declares that physics stops there; it does not permit us to look beyond.

Through black holes, general relativity opened up other unforeseen vistas. The first black hole solution to Einstein's equation was discovered in 1916 by the German astrophysicist Karl Schwarzschild, while he was serving on the front lines during the First World War. However, acceptance of its physical meaning came very slowly. A natural avenue for the formation of black holes is stellar collapse. While stars shine by burning their nuclear fuel, the outward radiation pressure can balance the inward gravitational pull. But after the fuel is all burned out, the only known force that can combat gravitational attraction comes from the quantum mechanical Pauli exclusion principle. During his celebrated voyage to Cambridge, the 20-year-old Subrahmanyan Chandrasekhar combined principles of special relativity and quantum mechanics to show that, if a star is sufficiently massive, gravity would overwhelm the Pauli repulsive force. During the 1930s, he refined his calculations, providing irrefutable arguments for the stellar collapse. However, the leading pre-eminent British astrophysicist of the time, Arthur Eddington, abhorred the idea of stellar collapse and declared that in the 'correct' calculation, special relativity had to be abandoned![1] This delayed not only the recognition of Chandrasekhar's work, but also the general acceptance of black holes by several decades.

Ironically, even Einstein resisted black holes. As late as 1939, he published a paper in the *Annals of Mathematics* arguing that black holes could not be formed by the gravitational collapse of a star. The calculation is correct, but the conclusion is an artifact of a non-realistic assumption. Just a few months later, American physicists Robert Oppenheimer and Hartland Snyder published their classic paper establishing that black holes do in fact result. These are regions in which the space-time curvature is so strong that even light cannot escape. Therefore, according to general relativity, to outside observers they appear pitch black. In the rubber sheet analogy, the bending of space-time is so extreme inside a black hole that space-time is *torn-apart*, forming a singularity. As at the big-bang, curvature becomes infinite. Space-time develops a final boundary and physics of general relativity simply stops.

And yet, black holes appear to be mundanely common in the universe. General relativity, combined with our knowledge of stellar evolution, predicts that there should be plenty of black holes with ten to fifty solar masses, the end products of the lives of large stars. Indeed, black holes are *prominent* players in modern astronomy. They provide powerful engines for the most energetic phenomena in the universe such as the celebrated gamma ray bursts in which an explosion spews out, in a few blinding seconds, as much energy as a 1000 suns do in *their entire lifetime.* One such burst is seen every day. Centres of all elliptical galaxies appear to contain huge black holes of millions of solar masses. Our own galaxy, the Milky Way, has a black hole of about 3.2 million solar masses at its centre.

BEYOND EINSTEIN

A really new field of experience will always lead to crystallization of a new system of scientific concepts and laws. When faced with essentially new intellectual challenges, we continually follow the example of Columbus who possessed the courage to leave the known world in an almost insane hope of finding land again beyond the sea.

W. Heisenberg, 'Recent Changes in the Foundation of Exact Science'

General relativity is the best theory of gravitation and space-time structure we have today. It can account for a truly impressive array of phenomena [1, 2] ranging from the grand cosmic expansion to the functioning of a mundane global positioning system on earth. But it is incomplete because it ignores quantum effects that govern the sub-atomic world. Moreover, the two theories are dramatically different. The world of general relativity has geometric precision, it is deterministic; the world of quantum physics is dictated by fundamental uncertainties, it is probabilistic. Physicists maintain a happy, schizophrenic attitude, using general relativity to describe the large-scale phenomena of astronomy and cosmology and quantum mechanics to account for properties of atoms and elementary particles. This is a viable strategy because the two worlds rarely meet. Nonetheless, from a conceptual standpoint, this is highly unsatisfactory. As physicists tells us that there should be a grander, more complete theory from which general relativity and quantum physics arise as special, limiting cases. This would be the quantum theory of gravity. It would take us beyond Einstein.[2]

Nevertheless, due to the inner atomic movement of electrons, atoms would have to radiate not only electromagnetic but also gravitational energy, if only in tiny amounts. As this is hardly true in Nature, it appears that quantum theory would have to modify not only Maxwellian electrodynamics but also the new theory of gravitation.

At the big-bang and black hole singularities the world of the very large and of the very small meet. Although they seem arcane notions at first, these singularities are our gates to go beyond general relativity. It is now widely believed that real physics cannot stop there, rather, general relativity fails. We need to dramatically revise, once again, our notions of space and time. We need a new syntax.

Creation of this syntax is widely regarded as the greatest and the most fascinating challenge faced by fundamental physics today. There are several approaches. While they generally agree on a broad list of goals, each focuses on one or two features as the central ones, to be resolved first, in the hope that the other problems 'would take care of themselves' once the 'core' is well-understood. Here, I will focus on loop quantum gravity which originated in our group some twenty years ago and has been developed by about two dozen groups worldwide (Ashtekar 2000). It is widely regarded as one of the two leading approaches, the other being string theory (Wadia).

In general relativity, space-time is modelled by a continuum. The new idea is that this is only an approximation, which would break down at the so called Planck scale—the unique length, $l_{pl} = \sqrt{G\hbar/c^3} \sim 10^{33}$ cm, that can be constructed from Newton's constant of gravitation G, Planck's constant $\hbar$ of quantum physics, and the speed of light c. This scale is truly minute, some 20 orders of magnitude smaller than the radius of a proton.[3] Therefore, even in the highest energy particle accelerators on earth, one can safely work with a continuum. But the approximation would break down in more extreme situations, in particular, near the big-bang and inside black holes. There, one must use a quantum space-time of loop quantum gravity.

What is quantum space-time? Look at the sheet of paper in front of you. For all practical purposes, it seems continuous. Yet we know that it is made of atoms. It has a discrete structure which becomes manifest only if you zero-in using, say, an electron microscope. Now, Einstein taught us that geometry is also a physical entity, on par with matter. Therefore, it should also have an atomic structure. To unravel it, in the mid-1990s researchers combined the principles of general relativity with quantum physics to develop a *quantum theory of geometry*. Just as differential geometry provides the mathematical language to formulate and analyse general relativity, quantum geometry provides the mathematical tools and physical concepts to describe quantum space-times (Ashtekar 2000, forthcoming).

In quantum geometry, the primary objects—the fundamental excitations of geometry—are one-dimensional. Just as a piece of cloth appears to be a smooth, two-dimensional continuum although it is obviously woven by one-dimensional threads, physical space appears as a three-dimensional continuum, although it is in fact a coherent superposition of these one-dimensional excitations. Intuitively, then, these fundamental excitations can be thought of as *quantum threads* which can be woven to create the fabric of space-time. What happens, then, near space-time singularities? There, the continuum approximation fails. The quantum fluctuations are so huge that quantum threads can no longer be frozen into a coherent superposition. The fabric of space-time is ruptured. Continuum physics rooted in this fabric stops. But the quantum threads are still meaningful. Using a quantum generalization of Einstein's equations one can still do physics, describe what happens in the quantum world. In the absence of a space-time continuum, many of the notions habitually used in physics are no longer available. New concepts have to be introduced, new physical intuition has to be honed. In this adventure, quantum Einstein's equations pave the way.

Using these equations recently the big-bang has been analysed in some detail (see, Ashtekar [forthcoming]). It turns out that the partial differential equations of Einstein's, adapted to the continuum, have to be replaced by difference equations, adapted to the discrete structures of quantum geometry. Except very near the big-bang, equations of general relativity provide an excellent approximation to the more fundamental ones. However, the approximation breaks down completely near the big-bang, when the density ρ of matter approaches the Planck density $\rho_p i = c^3/G^2\hbar \sim 10^{94}$gm/cc. In quantum geometry, space-time curvature does become very large in this *Planck regime,* but not infinite. Very surprisingly, quantum geometry effects give rise to a new *repulsive* force, which is so strong that it overwhelms the usual gravitational attraction. General relativity breaks down. The universe bounces back. But quantum Einstein's equations enable us to evolve the quantum state of geometry and matter right through this Planck regime. The big-bang is replaced by a quantum bounce.

Reliable numerical calculations have been performed in the strict spatially homogeneous isotropic case. Continuum turns out to be a good approximation outside the Planck regime also on the 'other side of the big-bang' (Ashtekar 2000, forthcoming). More precisely, in a forward-in-time motion picture of the universe, there is a contracting pre-big-bang branch well described by general relativity. However, when the matter density is approximately $0.8\rho_{pl}$, the repulsive force of quantum geometry, which is negligible until then, now becomes dominant. Instead of continuing the contraction into a big-crunch, the universe undergoes a big bounce, joining on to the post-big-bang expanding branch we now live in. Classical general relativity describes both branches very well, except in the deep Planck

Fig. 7.2 An artist's representation of the extended space-time of loop quantum cosmology. Time again runs vertically. General relativity provides only the top half of this space-time which originates in the big-bang. Quantum Einstein's equations extend this space-time to the past of the big-bang. The pre-big-bang branch is contracting and the current post-big-bang branch is expanding. The band in the middle represents the 'quantum bridge' which joins the two branches and provides a deterministic evolution across the 'deep Planck regime' (Courtesy Dr Cliff Pickover, www.pickover.com).

regime. There the two branches are joined by a quantum bridge, governed by quantum geometry.

The emergence of a new repulsive, quantum force has a curious similarity with the repulsive force in the stellar collapse as mentioned above. There, a repulsive force comes into play when the core approaches a critical density, $\rho_{\text{crit}} \approx 6 \times 10^{16}$ gms/cc, and can halt further collapse, leading to stable neutron stars. This force, with its origin in the Pauli exclusion principle, is *associated with the quantum nature of matter.* However, as indicated in section 'Big-Bang and Black Holes', if the total mass of the star is larger than, say, five solar masses, classical gravity overwhelms this force. The *quantum geometry repulsive force* comes into play at *much* higher densities. But then it overwhelms the standard gravitational attraction, *irrespective of how massive the collapsing body is.* Indeed, the body could be the whole universe! The perspective of loop quantum gravity is that it is this effect that prevents the formation of singularities in the real world, extending the 'life' of space-time through a quantum bridge.

Currently, work is under way to extend these results to more and more sophisticated models which incorporate inhomogeneities of the present-day universe. If the above scenario turns out to be robust, there will be fascinating philosophical implications for the issue of the Beginning and the End. For, the very paradigm to pose questions will again be shifted. If the questions refer to the notion of time that Einstein gave us, there was indeed a Beginning. Not at the big-bang though, but 'a little later' when space-time can be modelled as a continuum. But if by Beginning one means a firm boundary beyond which physical predictions are impossible, then the answer is very different from that given by general relativity: in the more complete theory, there is no such Beginning.

To summarize then, thanks to Einstein, our understanding of space and time underwent a dramatic revision in the twentieth century. Geometry suddenly became a physical entity, like matter. This opened up entirely new vistas in cosmology and astronomy. But a new paradigm shift awaits us again in the 21st century. Thanks to quantum geometry, the big-bang and black hole singularities are no longer final frontiers. The physical, *quantum* space-time is much larger than what general relativity had us believe. The existence of these new and potentially vast unforeseen domains has already provided a fresh avenue to resolve several longstanding, problems concerning both cosmology and black holes in fundamental physics. Even more exciting opportunities arise from new questions and the rich possibilities that this extension opens up.

NOTES

1. Today, a Ph.D. student would fail his qualifying exam if he/she were to make such an argument. Leading quantum physicists like Bohr and Dirac readily agreed with Chandrasekhar, but did not think it was worthwhile to point out Eddington's error publicly. It was only in 1983 that Chandrasekhar was awarded the Nobel Prize for this seminal discovery (Wali 1990).

2. Contrary to the common belief—rooted in Einstein's later views on incompleteness of quantum mechanics— he was quite aware of this limitation of general relativity. Remarkably, he pointed out the necessity of a quantum theory of gravity already in 1916! In a paper *Preussische Akademie Sitzungsberichte* he wrote:

    Nevertheless, due to the inneratomic movement of elections, atoms would have to radiate not only electromagnetic but also gravitational energy, if only in tiny amounts. As this is hardly true in *Nature*, it appears that quantum theory would have to modify not only maxwellian electrodynamics but also the new theory of gravitation.

3. For non-experts, it is often difficult to imagine how large a number 10^{20} is. So, the following illustration may help: $\$10^{20}$ would suffice to cover the US budget for a 100 million years at the 2005 rate!

REFERENCES

Ashtekar, A. (ed.) (2005). *100 Years of Relativity. Space-time Structure: Einstein and Beyond.* Singapore: World Scientific.

——— (2000). *The Universe: Visions and Perspective,* in N. Dadhich and A. Kembhavi (eds). Dodrecht: Kluwer Academic.

——— (forthcoming). In *Einstein and the changing world view of physics.* Berlin: Springer.

Chandrasekhar, S. (1991). *Truth and Beauty. Aesthetics and Motivations in Science.* New Delhi: Penguin Books, India.

Wadia, S. (ed.) (2005). Special Section, 'The Legacy of Albert Einstein', *Current Science*, 89, 12 2034‑2046.

Wali, K.C. (1990). *Chandra, A Biography of S. Chandrasekhar*. New Delhi: Penguin Books.

Will, C. (1986). *Was Einstein Right?*. New York: Basic Books Inc.

8

NARESH DADHICH

Why Einstein (Had I been born in 1844!)?

Let me begin by hypothesizing that it is tempting to place Einstein or oneself in the pre-Maxwellian times and follow the natural line of thought, which is in the Einsteinian spirit, and see what happens. Further in the present context, what type of questions does this line of thought give rise to? In this chapter, I wish to follow the Einsteinian spirit of consistency of principle and concept, and let the rest follow naturally all by itself.

We shall begin by defining universal entities and then identifying their primary examples. The rest of the story is built on seeking a relation between the universal entities and universalization of concepts.

Let us begin at the very beginning by defining universality and the process of universalization. The natural definition for universal is that it is the same for all and shared by all. It could be an entity or concept like space, or could be a force like gravity. Since a universal entity is the same and shared by all, this means all universal things must be related. That is, no two universal things can be independent. Any feature that distinguishes one universal thing from the other will mean that there exists some property, which is not the same for all. This will violate the universal character. If there exist two universal entities or concepts, they must thus be related and the relation would have to be universal naturally.

PRIMARY UNIVERSAL ENTITIES

What are the most primary universal entities we know of? The natural and obvious answer is space and time. They are indeed the same and shared by all things that exist in nature. Two questions arise: since both are universal, hence one, they must be on the same footing and second, there must exist a universal relation between them.

We know that the distance between two points depends upon the path an observer takes in going from one point to the other point in space. It is a common experience that kilometre reading in a taxi is path-dependent and that is why we are quite watchful that the driver takes the shortest and not the circuitous path. Thus spatial distance is path-dependent and so must be the time interval between any two events. This is what would be required to bring the two universal entities, space and time, on the same footing. This is however not so in the familiar Newtonian world. If we are to adhere steadfast to our concept of universality, we are forced to seek a new framework.

Second, as argued above since both space and time are universal, there must exist a universal relation between them. The most natural relation between them is through velocity (which is formally defined by, velocity = space/time). Therefore, a universal velocity is required which is the same for all observers irrespective of their relative motion. In the Newtonian mechanics, velocities are added by the law $w = u + v$. For instance, let two cars have velocity u and v relative to you on the ground, the relative velocity between the cars will be $w = u + v$. Existence of a universal velocity (which is the same for all and hence if one of u and v is the universal velocity, c, then w must also be c) cannot be compatible with the Newtonian law of addition of velocity and consequently with the Newtonian mechanics. It therefore asks for a new mechanics.

Thus we need a new mechanics simply by appealing to the universal character of space and time. One of the natural consequences of the existence of universal velocity is that like the spatial distance between two points, the time interval between any two events will now become path-dependent as required for space and time to be on the same footing. Existence of universal velocity thus addresses both the above questions.

UNIVERSAL VELOCITY

What should characterize the universal velocity, c?

(i) An object moving with c should always be moving, never at rest relative to any observer.
(ii) It should be a limiting velocity for all observers, no observer can attain c.
(iii) Since an object moving with c can never be at rest, it cannot have non-zero (rest) mass. Its mass is zero.
(iv) Existence of universal velocity means existence of zero mass particle.

What physical phenomenon could provide universal velocity? To address this question, let us get at the root of the phenomenon of motion. It is of two kinds, particle-like and wave-like. A particle can be given arbitrary velocity, could be held at rest while a wave's motion is entirely determined by the elastic properties of the medium in which it propagates, it is always moving and can never be held at rest. A wave's velocity could thus be changed only either by changing or moving the medium. That means a wave propagating in a universal medium, which could neither be changed nor moved, will have universal velocity, the same for all observers.

What is the universal medium? Obviously, space free of all matter is called vacuum. Hence the universal velocity we are seeking could then only be provided by a wave propagating in free space (vacuum). Simply on the force of consistency of the concept, we thus make a profound prediction that there must exist a wave propagation in free space and it would have a universally constant velocity relative to all observers.

HAD I BEEN BORN IN 1844!

Had I been born in 1844, it could have been quite possible to follow this train of thought. Then in around 1870, I could have, as a young man of 26, as old as Einstein was in 1905, made the above profound prediction that there must exist a wave propagating in free space with a universal constant velocity. That would have been a good five years before Maxwell's electrodynamics. It would have been simply remarkable that a prediction made on the force of pure thought and concept comes actually five years later in the form of the electromagnetic wave of Maxwell's electromagnetic theory. Hertz then experimentally establishes its existence. Light turns out to be the most familiar example of this new wave and thus its velocity becomes a universal constant.

A great conjecture—I shudder to imagine what would have happened if the events did happen as projected. Forget me, let us place Einstein as a young man in 1870, what he was in 1905, he could have very well argued as we did above and would have come up with the profound prediction before Maxwell's theory. Had that happened it would have been a great display of the pinnacle of human thought and Einstein as its purest human manifestation!

However, things do not happen like this, but it is insightful to wonder and probe the potential of sheer thought and analysis at its most pristine and sublime.

NEW MECHANICS

A new mechanics could have thus been predicted by the general principle of universality that all universal entities be on the same footing and be related by a universal relation. This leads to existence of a universal constant velocity and the incorporation of this fact gives rise to a new mechanics, which goes by the name, Einstein's special relativity. The year 2005 was celebrated as the Year of Physics in due recognition of the fact that it marks the 100th anniversary of this monumental discovery. (Simultaneously with this, he also made two other important discoveries of the photoelectric effect, which was just good enough to fetch him the Nobel Prize, and of the Brownian motion. But in the absence of this, they would not have merited this universal acclaim.)

The message of the new mechanics is that space and time are synthesized into one four-dimensional entity, space-time. Further it leads to the synthesis of mass and energy through the famous equation, $E = mc^2$, which has become synonymous with Einstein and much to his anguish and horror also with the atom bomb. Since the velocity of light, which is an electromagnetic wave, is universal—the same for all observers, it is the limiting speed of communication between spatially separated events in space. It therefore defines the velocity of causation. Events then get classified into two classes, those which can be causally connected, and those which cannot be, for instance, light takes eight minutes to travel from the sun to the earth. The event, which occurred at this moment on the sun, can affect (be causally connected) events on the earth only after eight minutes have elapsed. That is, the events occurring on the earth between now and until eight minutes can have no connection (cannot be influenced by) with this event on the sun. Thus we have finite velocity of physical communication—causal relation (causality), which is provided by the universal character of velocity of light. The rest of the Einsteinian mechanics is straightforward.

Another way of motivating special relativity could have been universalization of mechanics for all particles, massive as well as massless (Dadhich 2001; Dadhich 2002). The existence of a massless particle would have as argued above provided the universal constant velocity. Once that happens, Special Relativity is inevitable.

NEW GRAVITY

The next natural question is to universalize gravity. That is, it acts on everything including massive as well as massless particles. A particle of light, called photon is an example of a zero-mass particle. The massless particle, photon, propagates always with the universal constant velocity, which cannot change. Photon's velocity in vacuum is constant—a universal constant and hence it can never change. On the other hand, action of a force on anything is measured only through change in its velocity (motion). Velocity could, however, remain constant for circular motion, but light does not always move in a circle, instead in the Newtonian theory it always moves in a straight line.

We have thus landed with a serious contradiction in principle. Since gravity is universal, it must act on the massless photon as well, yet its velocity must not change. How is it possible? In the classical Newtonian framework, it is impossible to reconcile these two opposing properties. What should we do? We have no other way except to expand the framework (Dadhich 2003). How do we do that?

When one is confronted with such a question of concept, it is only robust common sense that can show the way. Let me take an uneducated peasant as personification of uninhibited common sense. I ask him this question. After some pondering, he says that he cannot much appreciate the action of gravity on light, and asks back what would you want light to do in actuality to feel gravity? I reply that if light is grazing past a massive body, it should bend toward it as every other thing does. He breathes deep and hard, scratches at his beard, and then asks me to follow him, and takes me to the river, which flows behind the village. He picks up a piece of log and throws that into the river. It starts floating freely with the flow of the river.

Then we walk along the river, which suddenly bends, and so does the log. He smiles and inquires, did any force act on the log? No, I say, it is freely floating. But even then it bent with the river? He then triumphantly inquires, haven't you got the answer to your question? Dumb as I am, I say, No. He asks me where does your light float? In space, I say. Then what is the problem, he says, bend the space. Then it dawned on me that why can't gravity bend/curve space around the massive body, and all particles including massless photons propagate/float freely in the curved space. In fact curved space-time, as space and time have already been synthesized into one in special relativity.

Isn't this simply astounding?

What have we arrived at? A new theory of gravity where it can honestly be described by no other means than the space-time curvature itself. Gravity no longer remains a force but gets synthesized with the structure of space-time and it simply becomes a property of the space-time geometry. Its dynamics, in particular the inverse square law of gravity, should now follow from the space-time curvature. That it does, all by itself. As Riemann curvature satisfies the Bianchi differential identities, which on contraction yields the second rank symmetric tensor with vanishing divergence. This leads to the Einstein equation incorporating the Newtonian inverse square law. Nobody has to prescribe a law of gravity. We have thus discovered a new gravity—Einsteinian gravity, called by the name General Relativity.

Gravity distinguishes itself from all other forces by the remarkable feature of impregnating space-time itself by its own dynamics. It then ceases to be an external force. Gravitational field is fully described by the curvature of space-time geometry. Motion under gravity will now be geodesic (straight line) motion relative to the curved geometry of space-time. The geodetic motion should naturally include the Newtonian inverse square attractive pull, and in addition it should also have the effect of curvature of space. Light cannot feel the former, but, only feels the latter. Stronger the gravity, stronger would be the curvature it produces. As we make the field stronger by making larger and larger mass confined to a smaller and smaller region, it is conceivable that space gets so curved that light cannot propagate out, but, its orbit closes on itself around the massive body. When that happens a black hole is defined from which nothing can propagate out. Things can only fall in but nothing can come out—it defines a one-way membrane. This is the most remarkable and distinguishing prediction of Einsteinian gravity.

Let me make one point clear that despite all general relativity literature being full of the phrase 'bending of light' due to gravity, it is as wrong as saying that the sun goes around the earth. What bends is space and light freely propagates in it. It was bending of space and not of light that was for the first time measured by Eddington in the 1919 total solar eclipse. We do however measure space bending by measuring deflection of light from the straight path. We do, in fact, see the sun going around the earth, yet we argue that it is apparent because we are sitting on the earth. Similarly, light can only freely propagate in space and cannot bend, and hence its deflection is the measure of space bending.

It is the universality of gravity, which has first demonstrated the inadequacy of the Newtonian framework, and has then indicated the expansion of the framework by way of curving space-time such that the contradiction is resolved. That has led to the new Einsteinian gravity.

IN HOW MANY DIMENSIONS DOES GRAVITY LIVE?

Everything about a universal force has to be determined all by itself— self-determined. One has no freedom to prescribe anything. We have seen that it is the property of universality, which leads to space-time curvature for description of the universal force, and then of its own determined the equation of motion for the force—gravity. The next natural question that can arise for the universal force is, in how many dimensions should it live? The minimum dimensions required to define the curvature tensor are two, but then it is well-known that the full dynamics of gravity cannot be realized in two and three dimensions. We thus come to the usual four dimensions in which the matter fields sit in three space dimensions. Why should matter remain confined to three-space, because its dynamics is fully realized in $(3 + 1)$-space-time? Hence there is no compelling physical reason to lift it to higher dimension. Should gravity then too remain confined to $(3 + 1)$-space-time?

Gravity is a self-interactive force, and self-interaction can only be computed iteratively. Since the space-time metric is the analogue of gravitational potential, the first iteration of self-interaction should involve the square of its first derivative. This is, however, automatically included in the Einstein equation which follows from the Riemann curvature involving second derivative and square of first derivative. The question is, why should we stop at the first iteration, shouldn't we also go to next iterations? We have to get everything from the curvature tensor, should we square it? That will give higher powers of first derivative, but will square second derivative as well. That is no good because the highest order of derivative must occur linearly in the equation to admit unique solution. If that is not the case, we will have more than one equation and hence more than one solution. This property is called quasi-linearity. We should thus seek the most general action for gravity constructed from the Riemann curvature and its contractions, which leads to the second-order quasi-linear differential equation.

Can we have second-order iteration of self-interaction and yet retain the quasi-linear character of the equation? The answer is yes. There exists a specific combination of square of Riemann, Ricci and Ricci scalar, called the Gauss-Bonnet combination, which precisely does this. We should hence include the Gauss-Bonnet term, which makes non-zero contribution only in dimension higher than four. This means physical realization of the second iteration of self-interaction of gravity demands that it propagate in the extra fifth dimension. It cannot remain fully confined to the three-space and its self-interaction dynamics takes it to leak into the extra space dimension. This is an important conclusion, which follows purely from classical consideration (Dadhich 2004a; Dadhich 2004b). It is remarkable that the differential geometry does provide a non-linear (in Riemann) action yet yielding the quasi-linear equation.

Another motivation for extra dimension for gravity comes from the general principle of charge neutrality. A classical force should always be overall charge neutral like the electromagnetic force. All bodies down to an atom are overall electric charge neutral. For gravity, matter in any form defines charge, which has positive energy and hence positive polarity. Gravitational charge is therefore unipolar. How do we now attain charge neutrality? The only option is to make the field have a charge of opposite polarity—negative. That is why universal force has to be self-interactive and always attractive! This is the most direct and simplest way to see why gravity is always attractive. Unlike electric charge, the negative polarity charge is non-localized and distributed with the field all over the space. If you integrate all over the space, negative charge of the field will fully balance the positive charge of matter field. However, in the local neighbourhood around the matter distribution, there will be charge imbalance, over-dominance of positive charge, and hence the field has to propagate off the three-space locally. However, it should not be able to go deeper in the extra dimension because as it propagates larger and larger region of negative charge (field) will get included and hence its strength will diminish exponentially (Dadhich 2003). This means gravity essentially remains confined with massless mode having ground state on the three-space while propagation into the extra dimension is effectively through massive modes. Interestingly this is precisely the picture presented by the Randall-Sundrum brane world gravity (Randall and Sundrum 1999). Note that ours is purely a classical argument, which does not take a priori existence of extra dimension, but it is solely dictated by the dynamics of gravity. This is a very important difference.

The curvature of space-time is not purely a geometric entity, but it is the carrier of gravitational dynamics. So long as curved space-time is not isometrically embedded in higher dimensional flat space-time, it means that curvature does transmit non-trivial gravitational dynamics into the extra dimension and it cannot be flat. In general, a four-dimensional curved space-time is not isometrically embeddable in five-dimensional flat space-time unless it is conformally flat. Hence gravity cannot remain fully confined to four-space-time. This is purely a geometric argument for higher dimension. On the other hand, it has been shown that an arbitrary four-dimensional space-time can always be embedded in five-dimensional Einstein space (de Sitter or anti de Sitter, Dahia and Romero 2002). This again shows that once we hit conformally flat higher dimensional space-time, the iteration chain should terminate.

Does this iteration chain stop or does it go on indefinitely to higher and higher dimensions? It naturally stops at the second iteration. In higher dimensional bulk space-time, only gravity propagates while matter fields remain confined to the three-space and hence the space-time is isotropic and homogeneous. It is then the maximally symmetric space-time of constant curvature, which automatically solves the equation with the Gauss-Bonnet term and the Gauss-Bonnet parameter defining the constant—in the bulk. It will be an Einstein space, which is conformally flat with vanishing Weyl curvature, that means there exists no more any free gravity to propagate any further in higher dimension. Since Weyl curvature is zero, it can be isometrically embedded in higher six-dimensional flat space-time. The iteration chain thus stops at five-dimension. The important point to realize is that it is the property of gravity, which naturally leads to higher dimension.

HOW MANY BASIC FORCES SHOULD THERE BE IN NATURE?

Gravity is the unique universal force, which is characterized by the following two properties:

(i) Universal linkage, interaction with all particles, massive as well as massless
(ii) Long range, present everywhere

It is shared by all that exist physically and hence could be taken as the mother force. All other forces should arise out of it by relaxing these two properties (Dadhich 2001; Dadhich 2003). How many different possibilities exist to give other forces in nature?

There are naturally three possible options: one, relax (i) and retain (ii), two, retain (i) and relax (ii) and lastly relax both (i) and (ii). Including the mother, there could therefore exist only four basic forces in nature. This is a very simple unified view of all the four forces. Let us next see, do these possibilities conform with the forces we know of?

In general, long-range forces will be classical while short-range forces will be quantum and nucleus-bound. There would exist two of them in each category. The mother force, gravity, is a long-range, classical tensor field. As we have seen, it is described by the curvature of space-time. The other long-range classical force will arise on relaxing the property (i) of linkage to all. That means, this force will link to particles having a specific new parameter, charge. There should exist a new parameter other than mass/energy, which will characterize this force. Again appealing to the principle of overall charge neutrality, the new charge will have to be bipolar (unipolar field could be only one for it has to be described by space-time geometry). This suggests that it will have to be a vector gauge field. It is a long-range force and hence it should obey charge conservation, which will lead to the usual inverse square law. All this clearly identifies the force to be Maxwell's electromagnetic with electric charge being the distinguishing charge parameter. Is this, like gravity, unique? Its dynamics is fully determined and hence there is no sensible reason for any other force to exist having the same dynamics as the electromagnetic field. Yes, it is indeed like gravity, unique.

Long range is characterized by massless and chargeless propagator so that it can freely propagate everywhere with the universal constant velocity. Short range on the other hand will be characterized by either propagator being massive and/or charged, or by coupling being running rather than constant. Short-range forces will be quantum and will remain confined inside some region like the nucleus.

Now if we relax the property (ii) and retain (i); that is, the force is short-range but has universal linkage to all (massive) particles that can remain confined to short-range, nucleus. The propagator is massive and/or charged and it negotiates interaction between massive electrically charged as well as neutral particles. It could be the weak force which interacts with all massive particles including neutrinos (Das and Ferbel 1993). Since this force interacts only with massive particles, it is predicted that neutrino must have non-zero mass. It is a kind of complementary ([i] but not [ii]) to the electric force ([ii] but not [i]). In an appropriate space there should exist a duality relation between them and the electroweak unification is perhaps indicative of that.

Lastly, let us relax both the properties, neither linkage to all (i) nor long range (ii). It is complementary to gravity, which respects both the properties. It could be the strong force through which the smallest building blocks of matter, quarks interact with each other (Das and Ferbel 1933). For the strong force, though propagator is massless but coupling is running which increases with distance and vanishes as distance goes to zero. This feature is known as the asymptotic freedom. This property is dual of the asymptotic freedom of long-range forces, which become free at infinity.

Clearly, unlike the long-range forces, it is not possible to establish uniqueness of the short-range nuclear forces. If there exists a new force, this is the right place to look for it. Until we can establish their uniqueness, the question will remain open. The other question, which this way of analysis prompts, is that there should, in some appropriate space, exist duality based on the complementary character of the forces. That is between the electromagnetic and the weak force, and between the gravity and the strong force. They have complementary features. There are some suggestive indications. In the electroweak unification, which is though not complete, it is significant that the two symmetry groups can be combined together as product $SU(2)XU(1)$. On the other hand, there exists in the string theory, now the celebrated *AdS/CFT* (anti de Sitter, conformal field theory) correspondence (Maldacena 1998; Witten 1998). At the boundary of the *AdS* space-time there lives the conformal field theory of matter—quantum chromodynamics. So we have *AdS* as gravity and *CFT* as the strong force dynamics of matter. This shows that there is a deeper connection between gravity and strong force. *AdS/CFT* correspondence is, in fact, a reflection of the duality between the gravity (space-time) and the strong force (matter).

We would thus strongly argue for probing duality relations indicated by the complementary features of the forces. Hopefully, it may lead to some new insights. If nothing else, there emerges a unified picture of the four forces in a very simple manner. Unification of all the forces is the driving theme for the string theory which is also expected to give a quantum theory of gravity (Schwarz 2000). String theory begins with the two basic universal principles, the special relativity and the quantum principle, and the rest follows from an attempt to construct a consistent theory of matter. First, it takes off to twenty-six dimensions and then renormalizability of the theory brings it down to ten or eleven-dimensions. The pertinent question now is to come down to the four-dimensional space-time we live in. Unfortunately there is no unique way to do that general relativity appears along with plethora of other scalar fields in four dimensions as an effective theory in the low-energy limit.

SEEING BEYOND EINSTEIN

The other great discovery, which signalled the turn of the twentieth century, was that the microstructure of matter has quantum character, it is made of discrete tiny pieces. That is, matter is not continuous all the way down. As we make pieces smaller, the process of observing (probe) them will require smaller and smaller energy. Then there would occur the limiting situation where the energy of the object and the probe become comparable. It is again not reconcilable with the classical Newtonian framework and we need a new mechanics of quantum mechanics.

This situation is characterized by the property that the process of observation disturbs the object non-ignorably. That is, a certain amount of uncertainty in observation becomes inherent which has to be incorporated in mechanics. That is, physical parameters will have to have probabilistic meaning and interpretation. This realization is the key to quantum mechanics. The wave motion offers the simplest manifestation of it. On the one hand, it is characterized by the four-wave-vector and on the other it should, like any other thing, carry energy and momentum, which is given by the four-momentum-vector. It stands to reason that the two four-vectors characterizing the same wave motion should be related, and this relation should be universal for it should be true for any wave motion. This leads to the basic relation between energy and frequency, and three-momentum and three-wave-vector, for which we have to introduce a new universal constant, called the Planck's constant. This constant is also the measure of the inherent uncertainty in measurement. It is obvious that it is impossible to localize a wave to determine its position precisely without making its momentum uncertain. Both position and momentum cannot simultaneously be determined to arbitrary accuracy. Accuracy of the one is only attainable at the cost of the other. This is the basic quantum principle, known as the uncertainty principle. The important message of quantum theory is that whenever object and probe become energetically comparable, motion tends to be wave-like. It could truly be described by quantum mechanics. At the micro-scale, quantum mechanics thus becomes inevitable.

With the discovery of electromagnetic wave and light its most common visible example, it is a pertinent and valid question to ask how does it propagate in vacuum? Should vacuum be completely physically inert or should it have some physical properties? This is a very contentious issue that brings forth the memory of infamous aether. Howsoever complex and involved the question be, we have to address it.

Let us get our facts right, that is, what have been observationally established? One, an electromagnetic wave propagates in vacuum and second, one cannot measure any motion relative to vacuum, that is it cannot act as a reference frame. It is the latter property, which has been taken as evidence for the untenability of the existence of aether. The former would naturally ask for oscillation of some basic constituents of the medium, vacuum to make wave propagate in it. Thereby it will, similar to matter, demand some kind of microstructure—basic building blocks of space. If that does not happen, nothing can propagate in it, nor can it bend as required by the Einsteinian gravity. Not only matter has quantum microstructure, space will also have to have some kind of structure. This is also required for physical realization of the quantum fluctuations of vacuum. If it has no microstructure, what will fluctuate? That means, even space cannot retain its continuum character all the way down.

Does this property necessarily have to conflict with the other, that it cannot act as a reference frame? No, because space is universal and hence is everywhere, it cannot have anything of its own that could have a distinguishing character. Hence it cannot define a reference frame for any motion. This is similar to a situation that in the middle of the ocean, without introducing a flag-post external to the ocean it is impossible to define a reference point. The ocean does have a microstructure yet it cannot define a reference frame. Similarly, space itself cannot define a reference frame but that can't stop it from having physical microstructure. It is again the universality property of space, which makes it reference-inert, even though it has microstructure. Hence there is no conflict between the two properties.

Quantum character is universal and hence space must also, like matter, acquire quantum behaviour at micro level. This view is further reinforced by Einsteinian gravity in which space-time acquires the physical property that it can bend like any other material medium. Recall the famous Wheelerian pronouncement, 'Matter tells space how to curve and space tells matter how to move'. Since space and matter respond to each other physically, they should at a deeper level share the same property, or rather be on the same footing.

Again the principle of universality demands that space-time should have quantum structure at micro level. The prime question then is, what are the basic building blocks of space-time? Nobody knows, that is, indeed the most challenging question of the day. To address this question, we will have to find quantum theory of space-time itself. That is what is being done in the canonical approach to quantum gravity, called the loop quantum gravity (Ashtekar 2004; Ashtekar and Lawandowski 2004). The Einsteinian gravity is described by the curvature of space-time, which is continuum, but at the micro-level it can no longer retain its continuum character. It has to turn into discrete quanta and hence, we need a new theory of gravity, which takes us beyond Einstein, a quantum theory of space-time and gravity.

On the other hand, the quantum principle (uncertainty principle) is universal and hence it must in some way be related with the most primary entity, space-time. It should really be deduced as a property of space-time. Until that happens, quantum theory will, in principle and concept, remain incomplete. The completion of the quantum theory is also therefore asking for a new theory. That will perhaps come about by realizing the quantum character of space-time itself, that is, to formulate quantum mechanics in a quantum space-time.

Doesn't it sound crazy and formidable? That it is. That is why the question has remained open for almost a century, which is in good measure indicative of its complexity both in concept and technique. It is interesting to note that both gravity and quantum theory are pointing at quantum character of space-time and its incorporation is thus imperative for a new theory of gravity and quantum mechanics. That is the key and the most challenging question of the day. The two most prominent approaches addressing this fundamental question are the particle theory-based string theory and the general relativity-based canonical loop quantum gravity. There is, however, a long way to go before a complete theory of quantum space-time and gravity emerges.

OUTLOOK

The main theme of the chapter is to demonstrate the power and sheer simplicity of pure thought and robust common sense. It is a tantalizing hypothesis that had a young person in the 1860s wondered about things and argued as we have done above, it would have been quite possible to predict the existence of a wave propagating with universal constant velocity in vacuum, and consequently the new mechanics of special relativity as well as new gravity of general relativity. All this could have happened without any experiment challenging the existing theory. This was precisely what had happened for Einstein's discovery of general relativity which was purely principle and thought-driven. We are taking one more step forward in that direction. Conceptually, all that was required was available not only in the 1860s, but right from Newton's time.

Why do I then single out the 1860s? Because before that the prediction would have been too much ahead of its time, and hence would have had a stillbirth. Take the example of the Greek philosopher, Aristarchus who is believed to have proposed that the earth goes around the sun and not the other way round as early as in the second century BC. This discovery could not survive for want of a proper intellectual base and understanding, which came into existence only in the fifteenth century through the observation of planetary orbits. Then the time was ripe for Copernicus to make the monumental discovery. It is in this context that the 1860s attain significance. With Maxwell's theory of electromagnetism coming soon, the predicted wave would have been identified with the electromagnetic wave to be observed experimentally by Hertz. The stage was therefore well set for the profound prediction and discovery.

It is interesting to conjecture and wonder, but the hard fact of life is that scientific discoveries are seldom driven, with the honourable and unique exception of general relativity, by pure thought. They are essentially driven by contradiction between theory and observation, and the latter also depends upon the available technology for instruments. Truly, they are the products of the times—the prevailing scientific thought process. However, had it happened as we envision, it would have been an amazing feat.

At any rate, it is a wonderful and straightforward way of looking at things based purely on simple logic and robust common sense. It is indeed insightful to wonder and ponder over the fundamental questions, such as the number of dimensions of space-time and number of basic forces in nature. In a natural way, one can argue that gravity requires a minimum of four-dimensions for the full realization of its dynamics, and it also leaks into the extra dimension but cannot penetrate deep enough. Thus four-dimensions are necessary but not sufficient for gravity. Further, it is the unique universal force characterized by the properties; linkage to all and long range. By peeling off these properties, it is remarkable to have a unified view of all the four basic forces in nature and it also indicates why there are only four of them. The complementary features of the forces strongly suggest the relation of duality between gravity and the strong force, and electromagnetic and the weak force. This is one of the neat predictions of this way of analysis. Lastly, it is important to realize that it is the completion of both quantum theory as well as general relativity that ask for a new quantum theory of space-time. The really challenging task is to do both gravity and quantum theory not in continuum, but in quantum space-time.

REFERENCES

Ashtekar, A. (2005). 'Gravity and the Quantum', *New Journal of Physics*, 7, 198, gr-qc/0410054.

Ashtekar, A. and J. Lawandowski (2004). 'Class. Quant. Grav.', 21, R53.

Dadhich, N. (2001). 'Subtle is the Gravity', gr-qc/0102009.

——— (2002). 'The Relativistic World: A Common Sense Perspective', physics/0203004.

——— (2003). 'Universalization as a physical guiding principle', gr-qc/0311028.

——— (2004a). 'Probing Universality of Gravity', gr-qc/0407003.

——— (2004b). 'Universality, Gravity, the enigmatic Lambda and Beyond', gr-qc/0405115.

Dahia, F. and C. Romero (2002). 'The embedding of the space–time in five dimensions: An extension of the Campbell-Magaard theorem', *J. Math. Phys.*, 43, 5804.

Das, A. and T. Ferbel (1993). *Introduction to Nuclear and Particle Physics*. United Kingdom: John Wiley.

Maldacena, J.M. (1998). 'The Large N Limit of Superconformal Field Theories and Supergravity', *Adv. Theor. Math. Phys.* 2, 231.

Randall, L. and R. Sundrum (1999). 'An Alternative to Compactification', *Phys. Rev. Lett.*, 83, 4690.

Witten, E. (1998). 'Anti De Sitter Space And Holography', *Adv. Theor. Math. Phys.* 2, 253.

Schwarz, J.H. (2000). 'Introduction to Superstring Theory', hep-ex/0008017.

9

B.N. JAGATAP

Bose-Einstein Condensation
When Atoms Become Waves

Bose-Einstein Condensation (BEC) is perhaps the shortest chapter in the life and science of Albert Einstein. The chapter begins with a letter written by Satyendra Nath Bose in June 1924 in connection with his paper on new derivation of Planck's Law (Bose 1924) and ends with the publication of Einstein's work by early 1925 (Einstein 1924, 1925). The essence of this work was the prediction that a gas of non-interacting bosons—particles with integral quantum spin, when cooled below a certain critical temperature would collapse into the lowest energy quantum mechanical state. The result would be an entirely new state of matter, where each particle would behave exactly like all the others. Einstein, thus, arrived at the first purely statistically derived example of a phase transition, which is now known as the BEC.

For a long time BEC was considered as theoretically fascinating, but experimentally impossible, owing to the extremely low temperatures required for its manifestation. Worse still is that at low temperatures the matter would have to be either solid or liquid, and BEC could only occur in a gas! It appears that even Einstein was not fully convinced for he wrote to Ehrenfest, 'From a certain temperature on, the molecules "condense" without attractive forces, that is, they accumulate at zero velocity. The theory is pretty, but is there also some truth about it?' (Pais 1982). In 1938, London proposed interpreting the HeI-HeII phase transition at 2.19 K as a BEC (1938). However, liquid helium is far removed from the ideal gas that was envisaged by Einstein and many features associated with BEC are masked by

---

[*] I thank Mr Piyush Pandey, Director, Nehru Planetarium, for his kind invitation to deliver this talk in the Einstein Lecture Series. The inspiration behind the title of the lecture is undoubtedly the Nobel Lecture of Prof. Ketterle (2002). I am indebted to Dr P.R.K. Rao for exciting discussions on Einstein and for his gift of the book The Science and the Life of Albert Einstein by Abraham Pais (1982). This book has been a source of most of the historical background of BEC in this article. Finally I thank my colleagues in the ultra-cold laboratory of BARC, Dr K.G. Manohar and Dr S. Pradhan, for many stimulating discussions on cold atoms and BEC.

by the strong interactions associated with this system (Griffin 1993; Griffin, Snoke, and Stringari 1995). Nevertheless the connection with the superfluid helium established BEC as a key element of macroscopic quantum phenomenon such as superfluidity and superconductivity.

The quest for BEC finally ended in 1995 when three research groups from USA successfully produced BEC in dilute gases of alkali atoms confined in magnetic traps and cooled by laser cooling and evaporative cooling techniques (Anderson et al. 1995; Davis et al. 1995; Bradley et al.). In a landmark experiment, C.E. Wieman and E.A. Cornell of Joint Institute for Laboratory Astrophysics (JILA) produced a condensate of ^{87}Rb (Anderson et al. 1995) at temperature of 170 nK (1nK = 10^{-9} K)[8]. Using different experimental configurations, W. Ketterle of Massachusetts Institute of Technology (MIT) and R.G. Hulet of Rice University reported observation of BEC in dilute atomic gases of ^{23}Na and ^{7}Li respectively (Davis et al. 1995; Bradley et al. 1995). The ground-breaking accomplishments of this research have been many—to cool neutral atoms to unprecedented low temperatures, thus exerting an ultimate control over their positions and velocities limited only by the uncertainty relation; to generate a sample of coherent matter wave consisting of atoms all occupying the same quantum state; to create a gaseous quantum fluid with properties that are different from helium quantum liquid; and to provide a versatile platform for precise testing of foundations of physics. Added to these are the excitements generated by the prospects of utilizing ultra-cold atoms and coherent matter waves in exotic applications such as quantum computation, gravity gradiometers, nano-fabrication, cold chemistry, etc. The field has already been awarded two Nobel prizes in physics—to S. Chu, W.D. Phillips, and C. Cohen-Tannoudji in 1997 for the development of laser cooling of atoms and to C.E. Wieman, E.A. Cornell, and W. Ketterle in 2001 for demonstration of BEC in alkali atomic vapours. This article attempts to provide a simple account of the seventy long years of search for Bose-Einstein condensate—a state of matter where atoms behave like waves (Ketterle 2002, Nobel lecture). Excitements generated by the research work in the very recent years and the promise that the field holds for tomorrow are briefly discussed here.

Historical

In June 1924, Einstein received a letter from Satyendra Nath Bose, a young lecturer at Dacca University. The letter was about a paper on new derivation of Planck's Law by Bose, which was apparently rejected by a referee of the *Philosophical Magazine*. Bose sent a copy of this paper written in English and requested Einstein to arrange for publication in the *Zeitschrift fur Physik*, if he felt the work worth publishing. Bose wrote, 'Respected Sir, I have ventured to send you the accompanying article for your perusal. I am anxious to know what you think of it. You will see that I have ventured to deduce the coefficient $8\pi\nu^2/c^3$ in Planck's Law independent of the classical electrodynamics' (Pais

1982). The expression $8\pi\nu^2/c^3$ arises in the derivation of Planck's formula through the number of states Z in the frequency interval between ν and $\nu + d\nu$ and Z is computed by counting the number of standing waves in a cavity of volume V to yield $Z = (8\pi\nu^2/c^3)\, V d\nu$. What Bose had done in his paper was to use the particle language for photons and to replace the counting of frequencies by the counting of cells in the one-particle phase space—First integrate the one-particle phase space element $d\vec{x}\, d\vec{p}$ over the volume V and over all momenta $\vec{p}$ and $\vec{p} + d\vec{p}$ to produce the quantity $4\pi V p^2 dp$. Then supply the relation, $p = h\nu/c$, where h is Planck constant, ν is the frequency of photon, and c is the velocity of light. Equate the quantity $4\pi V p^2 dp$ to $h^3 Z$ where Z is the number of cells of size h^3 contained in the phase space of the particle to obtain $Z = (4\pi\nu^2/c^3)\, V d\nu$. At this stage Bose took a bold step to introduce a factor of 2 to count the polarization of the particles, and the expression was exact. In the later part of the paper, Bose introduced his new statistics, which involved the indistinguishability of particles and the statistical independence of cells of the phase space to arrive at the Planck's Law.

Einstein found Bose's derivation beautiful as he wrote in a letter to Ehrenfest. The importance of the expression beautiful may be appreciated from what Einstein's son, Hans once said about his father, 'He had a character more like that of an artist than a scientist. The highest praise for a good theory or a good piece of work was not that it was correct not that it was exact but that it was beautiful' (Whittrow 1967). Einstein submitted Bose's paper to the journal after translating it in German and adding a translator's note: 'In my opinion, Bose's derivation of the Planck formula constitutes an important advance. The method used here also yields the quantum theory of the ideal gas, as I shall discuss elsewhere in more detail'(Bose 1924). Einstein adopted the Bose statistics and photon analogy to discuss the quantum theory of the ideal gas. The motivation to do this was simple enough, 'If it is justified to conceive of radiation as a quantum gas then the analogy between the quantum gas and a molecular gas must be complete'(Einstein 1925). To do so, however, Einstein needed Z appropriate for non-relativistic particles with mass and a constraint to conserve the number of particles, unlike Bose's derivation where the photon number was not conserved. Most of this work is covered in two papers, one published in 1924 (Einstein 1924) and the second in early 1925 (Einstein 1925), which brings forth the idea of Bose-Einstein condensation. When the ideal quantum gas is cooled below the critical temperature, Einstein notes (1925) that ' ... in this case, a number of molecules steadily growing with increasing density goes over in the first quantum state (which has zero kinetic energy) ... . A separation is affected; one part condenses, the rest remains a saturated ideal gas.'

BOSE-EINSTEIN CONDENSATION
Bosons and Fermions

The particles occurring in nature have intrinsic spin (I) which can have either integral, that is, 0, 1, 2, ... or half integral, that is, 1/2, 3/2, 5/2 ... values and accordingly called as bosons or fermions. The familiar example of bosons is a photon ($I=1$) while that of fermions is an electron or a proton, each having $I=1/2$. An atom also has a spin that is determined by the sum of spins of the electron and the nucleus constituting the atom. Hydrogen atom is a boson—it consists of an electron and a proton, and their spins add to an integral value. Alkali atoms ^{23}Na and ^{87}Rb are bosons, since their nuclear spins are respectively 5/2 and 7/2 while the unpaired electron in their outermost shell has a spin 1/2. In the similar manner ^{7}Li is a boson whereas ^{6}Li is a fermion. In general an atom consisting of even or odd number of particles, (electrons + protons + neutrons), is a boson or a fermion.

At the microscopic level there are profound differences between bosons and fermions. Bosons prefer to live together in a sense that they have no restriction on how many particles can be in any single particle state. Fermions on the other hand choose to avoid each other and as a consequence no more than one particle can be in the same state. This difference is illustrated in Fig. 9.1 and is summarized by the statement that the bosons obey the Bose-Einstein statistics and the fermions follow the Fermi-Dirac statistics (Liftshitz and Pitaevskii 1980). Also the fundamental quantum mechanical symmetry requirement is that the total wave function for a collection of bosons (fermions) is symmetric (anti-symmetric) with respect to the interchange of any two particles.

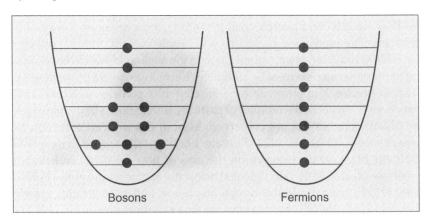

Fig. 9.1   Occupation of single particle states by bosons and fermions.

At normal or high temperatures a dilute gas of atoms behaves classically and therefore we do not observe any perceptible differences in the behaviour of a Bose gas and a Fermi gas. The effect of quantum statistics becomes evident at low temperatures when the gas starts to become a 'quantum soup' of indistinguishable particles. It is here that a Bose gas and a Fermi gas begin to behave very differently. A Bose gas undergoes BEC at a precise temperature with appearance of a condensate where all atoms occupy the same quantum state. A Fermi gas develops into a 'Fermi sea' wherein exactly one atom occupies each low energy state.

BEC as a Phase Transition

Consider an ideal Bose gas of N non-interacting particles of mass M contained in a box of volume V. The eigenstates of a single particle in a box are states $|k\rangle$ associated with the momentum $|p_k| = \hbar k$ where $\hbar = h/2\pi$. The statistical description of the gas is given by specifying the occupation number N_k—the number of particles in a state $|k\rangle$. For a particle we can define the de Broglie wavelength $\Lambda = \hbar/p$. For a gas at temperature T, the momentum p can be obtained from the average kinetic energy and that yields $\Lambda = (2\pi\hbar^2/Mk_BT)^{1/2}$ where k_B is the Boltzmann constant. At normal temperature, Λ is of the order of a fraction of a nm (10^{-9} m); however, it can be increased by decreasing the temperature of the gas. The number density of the particles in the gas is $n = N/V$ and consequently the mean inter-particle separation is $d \sim n^{-1/3}$. When $\Lambda << d$, the quantum effects are negligible, the particles behave classically and the properties of the gas are dominated by the thermal motion. On lowering T, Λ begins to approach d and the effects of quantum statistics become evident. In the case of bosons, which have no restriction on occupancy of a state, the quantum behaviour is manifested by an increase in the occupation of states $|k\rangle$ with small momenta. As the temperature is reduced further we attain $\Lambda \sim d$. At this stage the quantum statistical effects become dominant and the gas is said to become quantum degenerate. Eventually a macroscopic occupation of $p=0$ state develops when

$$n\Lambda^3 = n\left(\frac{2\pi\hbar^2}{Mk_BT}\right)^{3/2} \geq \zeta(3/2) = 2.612\ldots \qquad (1)$$

where ζ is the Riemann zeta function and $\zeta(3/2)$=2.612 ... This in essence is the BEC, which may be visualized as a phase transition where one phase, the Bose-Einstein condensate phase, is composed of particles with $p=0$ and the other phase consisting of particles with $p \neq 0$. The quantity $n\Lambda^3$ is called the dimensionless phase space density. The region given by eq. (1) is the condensation region and the critical temperature T_c is given by (Liftshitz and Pitaevskii 1980)

$$T_c = \frac{2\pi\hbar^2}{Mk_B}\left(\frac{n}{\zeta(3/2)}\right)^{2/3} \qquad (2)$$

For a given temperature we may also define a critical density n_c such that

$$n_c = \frac{\zeta(3/2)}{\Lambda^3} \qquad (3)$$

In terms of the condensate fraction N_0/N the thermodynamic phase change is described by

$$\frac{N_0}{N} \begin{cases} = 0, & T > T_c \\ = 1 - \left(\frac{T}{T_c}\right)^{3/2}, & T < T_c \end{cases} \quad (4)$$

Thus for $T > T_c$ no single state is occupied by a finite fraction of all the particles, whereas for $T < T_c$ a significant fraction occupies the level $p = 0$ while the rest of the particles are 'spread thinly' over the levels with $p \neq 0$. This is illustrated in Fig. 9.2. The abrupt occurrence of a finite occupation in a single state indicates a spontaneous change in the symmetry of the system and an ordering in the momentum space rather than in a real space. This makes BEC distinctly different from other phase transitions, for example, gas-liquid or liquid-solid. The condition $\Lambda \sim d$ for BEC also suggests that the wavefunctions of individual particles overlap to generate a macroscopic coherent state (Ketterle 2002; Anglin and Ketterle 2002), which in essence is the Bose-Einstein condensate. Experimental realization of such a state requires high atom number densities and/or extremely low temperatures. These stringent requirements may be easily appreciated by considering the example of hydrogen atom for which eq. (2) yields $n^{2/3}/T_c \sim 6.3 \times 10^{13} \text{cm}^{-1} \text{K}^{-1}$. This implies that at $n = 10^{15}$ atoms/cc, $T_c = 150$ μK, while at $n = 10^{12}$ atoms/cc, $T_c = 1.5$ μK (μK $= 10^{-6}$ K).

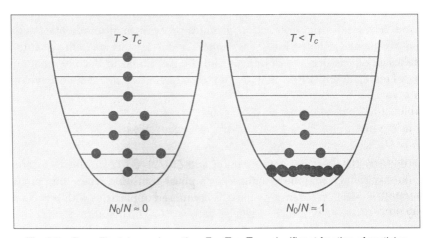

Fig. 9.2 Bose-Einstein condensation. For $T < T_c$, a significant fraction of particles occupy the lowest state of the potential confining the Bose gas.

BEC Analogies

BEC is an explicit quantum mechanical phenomenon and it is difficult to explain it in a simple form. The formal derivation of BEC found in the textbooks is generally too abstract to form a mental picture of the phenomenon. We may, however, consider a few analogous situations that may help develop a certain amount of physical insight about BEC. Consider first the example of a group of school children on a playground as illustrated in Fig. 9.3. We have chosen the children of equal height and wearing the school uniform so that they in a sense represent identical particles. The hands of each child are stretched and the extent of stretching is more like the de Broglie wave length Λ. We take the average separation between two children to be d which serves as the mean separation between the atoms in a gas. Fig. 9.3(a) describes the situation $\Lambda << d$ and corresponds to the normal state of a gas where thermal effects are dominant. Now, we organize the group in such a manner that $\Lambda >> d$. In the present example, this can be achieved by bringing the children closer and we may do so as long as the children enjoy being together just like the bosons. This situation is shown in Fig. 9.3(b). We observe that in this configuration the children become indistinguishable and the whole group as such develops coherence. This state is more like the Bose-Einstein condensate phase. There is one aspect, however, that we have not taken into account in Fig. 9.3(b). After all the children are children and there is a possibility that they keep kicking their adjacent neighbours as shown in Fig. 9.3(c). This kicking is analogous to the interaction between the particles of a gas. When the interaction is weak, the Bose-Einstein condensate state is stable. For strong interaction the Bose-Einstein condensate state becomes unstable and is eventually destroyed as depicted in Fig. 9.3(d). We thus learn that for observation of BEC the gas must be an ideal Bose gas, consisting of non-interacting or very weakly interacting bosonic particles.

Fig. 9.3 provides the simplest analogy to the phenomenon of BEC. It also makes it clear that the observation of coherence in a system is an indicator of BEC. This leads to further questions as to what causes the coherence to develop in a gas. The answer lies in the quantum statistical mechanics. Sacket and Hulet (2001) have developed a beautiful example to illustrate this aspect. They consider first a set of N 'Bose coins', which can be in one of the two states head (H) or tails (T) with equal probability. If the coins are classical distinguishable particles then the total number of possible configurations is 2^N. For example, if $N = 2$, the possible combinations are HH, HT, TH, and TT, which amounts to $2^2 = 4$. Note that the distinguishability of coins requires that the outcomes HT and TH must be counted as separate configurations. Out of these 2^N distinct configurations, consider the configuration $HHHHHH...$, where all the coins are in the same state H, just like all atoms in the same state for a Bose condensate. The probability P_H of achieving this state for the classical distinguishable coins is one in 2^N, that is, $P_H = 2^{-N}$. Now consider the situation where the coins are identical bosons. For $N = 2$, we can have only three distinct configurations HH, HT ($= TH$), and TT. The indistinguishability of particles requires that the outcomes HT and TH to be counted as one. If we continue this game of tossing of coins then we observe that for identical coins the total number of outcomes is $(N + 1)$

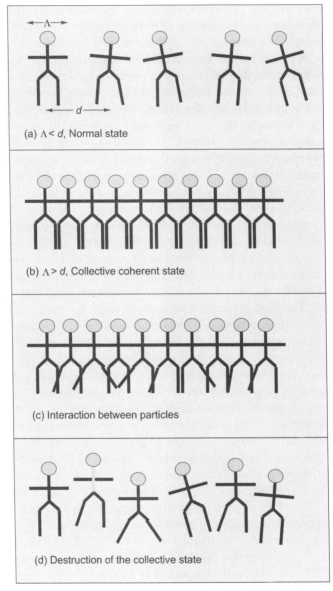

Fig. 9.3 Illustration of a simple analogy for Bose-Einstein condensation.

and the probability of having all the coins in H state is $P_H = (N+1)^{-1}$. Now if we take the limit $N \to \infty$, we find that $(N+1)^{-1} \gg 2^{-N}$, that is the probability of obtaining all the coins in the same state is incredibly larger for identical indistinguishable Bose coins in comparison to that for distinguishable coins. The message here is that a Bose condensate is formed for no other reason than that it is the most likely state of a sufficiently cold collection of bosonic atoms. This example of tossing of 'Bose' coins, however, does not explain the abrupt change in the state of the system that is so intrinsic to BEC. To do so Sackett and Hulet introduce the temperature dependence of the probability in the following manner (2001):

In a gas the probability of finding an atom in a state of energy E is given by the well-known relation $p = \exp(-E/k_B T)$. In the same spirit we consider the example of the 'Bose' coins where the probability of a single coin being in state H is p and that p is dependent on T. If we now continue the counting of states then we find that the relative probability of observing k coins in H is $p^k(1-p)^{N-k}$ and consequently the probability of observing all coins in H becomes

$$P_H = \frac{p^N}{\sum_k p^k(1-p)^{N-k}} = \frac{p^N(2p-1)}{p^{N+1}-(1-p)^{N+1}} \tag{5}$$

Fig. 9.4 shows a plot of P_H versus p for a few values of N. It is seen here that as N is increased there develops a discontinuity at $p = 0.5$. In fact, it can be shown that in the limit $N \to \infty$, eq. (4) takes the form

$$P_H = \begin{cases} 0 & \text{for } p < 0.5 \\ p^{-1}(2p-1) & \text{for } p > 0.5 \end{cases} \tag{6}$$

and the discontinuity at $p = 0.5$ in eq. (6) is analogous to the phase transition occurring in BEC.

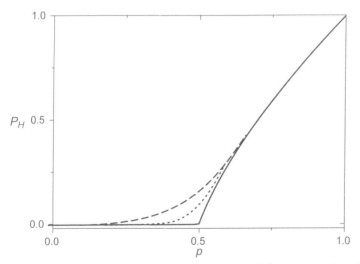

Fig. 9.4 Probability P_H of observing N coins in state H is plotted against the probability p that a single coin can be found in the state H. The curves corresponding to N = 2, 10, 1000 are shown by dashed, dotted, and solid lines respectively. The phase transition is evident when $N \to \infty$.

The Bose condensate is a coherent macroscopic state of identical bosons analogous to a laser, which is a coherent beam of photons. Conversely the normal state of the gas is like an ordinary incoherent light devoid of any phase correlation. In fact, the atoms in a Bose condensate formed in an atom trap exhibit several similarities with the coherent photons of a laser cavity. While these similarities come handy when discussing the 'atom laser', it must be remembered that BEC is clearly a thermodynamic process in contrast to a laser, which is an explicit non-equilibrium state due to the requirement of population inversion.

BEC: EXPERIMENTAL PERSPECTIVE

Systems for Observation of BEC

BEC is a phenomenon that occurs in physics at all scales, the fundamental requirement is that the particles be bosons. The systems range from gases, liquids, and solids and include semiconductors, metals, atomic nuclei, elementary particles, and astrophysical matter (Griffin 1995). Observation of coherence in such systems is an indicator to BEC. For example, Cooper pairs composed of electrons (e^-, e^-), or holes, (h^+, h^+) in metals or oxides exhibit coherence in superconductivity. The excitons (e^-, h^+) or biexcitons $2(e^-, h^+)$ are bosonic systems in semiconductors which display coherence in luminescence. In the high energy domain, the systems of interest are interacting bosons nn and pp in nuclei exhibiting coherence in excitation. Most of these systems are, however, strongly interacting like the case of liquid He and thus deviate severely from the ideal system envisaged by Einstein in his work on the quantum Bose gas.

There exists several difficulties associated with the experimental observation of BEC (Griffin 1995); no wonder BEC remained an unsolved problem in experimental physics till 1995. The problem begins with the identification of a Bose gas that remains dilute in terms of interactions, that is it must satisfy the condition $na^3 \ll 1$, where a is the scattering length which is related to the collisional cross-section $\sigma = 8\pi a^2$ for identical bosons and n is the particle number density as before. Hydrogen and alkali atoms have very small scattering lengths—$a = 0.072$ for hydrogen, ~ 5 nm for ^{87}Rb and ^{23}Na and -1.5 nm for ^{7}Li. The vapours or gases of these atoms provide ideal systems for observation of BEC. The gas chosen for BEC experiments must be contained in a manner that the three body collisions involving the walls of the container must be totally absent. This can be achieved by magnetic or optical traps, which provide the appropriate three-dimensional field boundaries for confinement of atoms—a truly wall-less confinement. The density of the gas must not be high enough to introduce collisions that lead to trap loss and heating of the sample. Working at moderate or low densities necessarily means that the gas must be cooled to extremely low temperatures, in the range of micro- and nano-Kelvin range, to reach the conditions required for quantum degeneracy that is $n\Lambda^3 > 1$. Over the years experimental physicists developed the techniques of 'evaporative cooling' (Griffin, Snoke, and Stringari 1995; Hess 1986; Gyeytak and Kleppner 1984; Silvera and Walraven 1986) and 'laser cooling' (Hansch and Schawlow 1975; Chu 1998; Cohen-Tannoudji 1998; Phillips 1998; Metcalf and van der Straten 1999) for cooling of atomic gases confined in optical and magnetic traps. A historical development of this research is provided in Table 9.1, which lists some of the major events in the pursuit of BEC of atomic gases.

Table 9.1 Bose-Einstein Condensation: Major Milestones

1924	Bose	Bose Statistics
1925	Einstein	Bose-Einstein Condensation
1938	London	Interpretation of Liquid ^{4}He
1982	Sears, Svenson, Martel, Woods	Condensate fraction ~ 0.1 in ^{4}He at $T = 0$

Spin Polarized Hydrogen (SPH) Atom Phase

1959	Hecht	SPH as a candidate for BEC.
1980	Silvera, Walraven	SPH at T = 270 mK, $n = 1.8 \times 10^{14}$
1985	Greytak, Kleppner	SPH at T = 300 mK, $n = 3 \times 10^{17}$
1986	Heiss	Idea of evaporative cooling
1986-98	Greytak, Kleppner Silvera, Walraven	T, n close to BEC transition point
1999	Greytak, Kleppner	BEC in hydrogen

Alkali Atom Phase

1975	Hansch, Schawlow	Idea of laser cooling
1984	Letokhove	Na atoms at 190 mK
1985	Phillips, Chu	100-50 mK in Na
1988	Phillips, Chu, Wieman	240 μK in Na, 100 μK in Cs
1990	Wieman, Phillips, Cohen-Tannoudji	1.1μK in Cs, 2 μK in He*
1991	Wieman, Pritchard, Ketterle	Hybrid approach
1995	Wieman, Cornell, Ketterle, Hullet	BEC in ^{87}Rb, ^{23}Na, ^{7}Li
1995-	Wieman, Cornell, Ketterle, Hullet	Properties of Bose Condensates
1997	Chu, Phillips, Cohen-Tannoudji	Nobel Prize for laser cooling
2001	Wieman, Cornell, Ketterle	Nobel Prize for BEC

Note: '*' indicates metastable atoms.

Evaporative Cooling of Atoms

Evaporative cooling as an efficient way of cooling atomic samples was first suggested in the context of cooling of spin-polarized hydrogen (Griffin, Snoke, and Stringari 1995; Hess 1986; Gyeytak and Kleppner 1984; Silvera and Walraven 1986). By definition evaporation is a process by which energetic particles leave the system with a finite energy thereby cooling the system to a lower temperature. Consider a system in thermal equilibrium at temperature T_1. This system inevitably contains particles of high energy distributed on the tail of the Maxwellian distribution. When these high energy particles escape from the system, they carry with them more energy than the average thermal energy of the system. On re-thermalization by the elastic collisions, the system attains a new distribution that corresponds to a temperature $T_2 < T_1$. This in essence is evaporative cooling.

Evaporative cooling of an atomic sample in a trap is schematically shown in Fig. 9.5. A trap is a device that confines a particle or collection of them to a region of space by a position dependent force. A trap can confine atoms for long duration of time required for manipulation and observations. It also prevents the atoms from reaching the physical walls of the experimental system, which is at much higher temperature. Fig. 9.5(a) provides an illustration of a sample of atoms confined by a magnetic field. The lowest barrier to escape the trap is represented by E_b. The distribution of atoms in the trap is such that atoms of lower energies lie at the bottom of the trap with smaller spatial extent and those with higher energies are distributed on the outer regions of the trap. Soon after the trap is loaded all atoms with energies greater than E_b are lost. The remaining atoms redistribute their energies through collisions and establish a new equilibrium corresponding to a lower temperature. This is 'spontaneous' evaporative cooling and is shown in Fig. 9.5(b). This process may be continued and atoms would continue to cool, but the process slows down as the temperature is reduced. Although total number of atoms decrease in this process, the density need not. In fact, the cooler atoms fall deeper in the trap and the density actually increases. Evaporative cooling is driven by the elastic collisions and they decide the re-thermalization time. For an efficient evaporative cooling the re-thermalization time scale must be shorter than the lifetime of the trapped sample.

Evaporative cooling may be forced by modifying in time the trap threshold to facilitate escape of less energetic atoms, as shown in the Fig. 9.5(c). The steady state temperature is proportional to E_b and consequently the gas can be cooled further by lowering E_b. The rate of change of E_b must be slow compared to the rate at which the atoms in the trap attain equilibrium. Another way of forcing an evaporative cooling is shown in (d). Here *RF* field is used to remove hot atoms from the trap in an energy selective way. In a magnetic trap the magnetic field increases with distance from the trap centre. The hot atoms can reach the outer regions of the trap and therefore experience higher magnetic field. In short the Zeeman shift becomes proportional to the energy of the atoms. An applied *RF* field then can flip the spin of the energetic particle in a selective manner and effect a transition from 'trapped' state to 'un-trapped' state. Thus only the higher energy atoms are removed from the trap by a careful choice of the *RF* field. This method of evaporative cooling offers many advantages. The magnetic potential is not required to be modified for evaporation and can be kept at the optimum confinement. The evaporation rate and energy carried out by the escaping particles can be easily controlled by the amplitude and frequency of the applied radiation. These

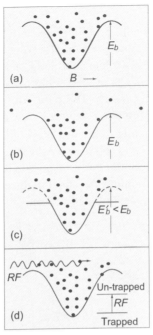

Fig. 9.5 Evaporative cooling of atoms in a magnetic trap. (a) Atoms are loaded in a trap. B is magnetic field and E_b is the trap threshold. (b) Spontaneous cooling where the atoms having energy greater than E_b escape the trap. (c) Forced evaporative cooling by changing the trap threshold from E_b to E'_b. (d) RF evaporative cooling where an RF radiation is used to release hot atoms from the trap in energy and space selective manner.

and other advantages make RF-evaporative cooling a technique of choice for experiments for observation of BEC (Anderson et al. 1995; Davis et al. 1995; Bradley et al. 1995).

The magnetic traps used for evaporative cooling are based on the force $F = -\nabla(\vec{\mu} \cdot \vec{B})$ where μ is the magnetic moment of the atom and B is the magnetic field. The simplest of the magnetic traps is the quadrupole trap where $|B|$ varies linearly in space and can be formed by two identical coils separated by a distance and carrying opposite currents. This trap has a centre where the field is zero and the field increases in all directions. An atom crossing the trap centre, however, can find its spin flipped and as a consequence it is ejected out of the trap. This makes a quadrupole trap leaky as if it has a hole at the zero field. This hole can be plugged by adding a small transverse magnetic field (Anderson et al. 1995) or using an 'optical plug' generated by a repulsive dipole force (Davis et al. 1995). Other examples of magnetic traps include Ioffe-Pritchard trap or a clover leaf trap (Metcalf and van der Straten 1999).

Laser Cooling of Atoms

Laser cooling refers to cooling of free atoms in a gas or vapour phase using laser beams that are tuned close to the atomic resonance (Hansch and Schawlow 1975; Chu 1998; Cohen-Tannoudji 1998; Phillips 1998; Metcalf and van der Straten 1999). The process relies on the momentum transfer in the laser-atom interaction process. The basic idea underlying laser cooling is illustrated in Fig. 9.6. Here we consider a moving atom illuminated by a laser beam that is propagating in the direction opposite to the velocity $\vec{v}$ of the atom. The frequency ν_L of the laser is tuned near a resonance line of the atom. On absorption of a photon the atom goes to the excited level and at the same time it receives momentum Δp equal to the photon momentum, hc/ν, in the direction of the propagation of the laser.

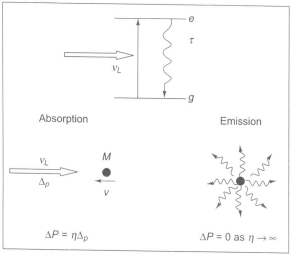

Fig. 9.6 Principle of laser cooling of atoms. A laser of frequency ν_L is directed to oppose the velocity (v) of the atom of mass M. The ground and excited levels of the atom are denoted by g and e respectively, and t stands for the lifetime of the excited level. Momentum change per photon absorption/emission is Δ_p. After a large number (η) of absorption emission cycles, the momentum change in the absorption process accumulates while that in the emission process averages out to be zero. The result is a force that is in the direction opposite to v.

Since the excited level has a finite lifetime ($\tau \sim 10^{-9}$ s), the atom decays back to the ground level by spontaneous emission of a photon. The emission process also involves change in the momentum of the atom, but this change is not in any preferred direction since the spontaneous emission is isotropic in nature. Now if we consider a large number (η) of photon absorption emission cycles then we find that the total momentum change due to absorption accumulates to $\Delta P = \eta \Delta p = \eta hc/\nu$, while the momentum change due to emission averages out to zero due to random directions of the emitted photons. The net result is that the atom gains momentum ΔP in the direction of the laser beam, which may be viewed as if the atoms are acted upon by a force. This force is called as the 'scattering force' and it can be used to decelerate the atom and thereby reduce its velocity and temperature. We illustrate this idea with a simple example of sodium atom. We take the laser beam to be near resonant to $3s \to 3p$ transition at 580 nm. During absorption of a photon, the change in the velocity of the atom is $\Delta v = \Delta p/M \sim 3$ cm/s. Sodium atoms produced at temperature of 500 K have an average thermal velocity of $v = 6 \times 10^4$ cm/s. In order to bring these atoms to rest, that is $v \to 0$, $T \to 0$, we need $\eta = v/\Delta v \sim 2 \times 10^4$, which is a very large number indeed. This, however, is not a matter of much concern since the lifetime (τ) of the excited $3p$ level of sodium is ~ 16 ns which is an average time taken for one cycle of photon absorption and emission. We thus see that the cooling time is $n\tau \sim 300$ μs, which is very small.

The scattering force discussed above can be used to decelerate a beam of neutral atoms by apposing its velocity by a laser beam. For a gas of atoms, a single laser beam is not sufficient. This is where Hansch and Schawlow (1975) noted that if the atoms were irradiated by a set of counter propagating beams tuned to the lower frequency side (redshifted) of the atomic absorption line, there would be a net force opposing the velocity of the atoms. In this configuration Doppler shift makes it more likely that the atoms absorb photon from the laser beam, which is nearly opposing its velocity. For example, an atom moving with a velocity $+ v_x$ will blueshift into resonance with a laser beam propagating along $-x$ direction and redshift out of resonance with the laser beam propagating along $+x$ direction. Consequently this atom is more likely to absorb photons from the laser beam directed towards $-x$ direction and experience scattering force in that direction. When the gas is irradiated with six laser beams along $\pm\hat{x}$, $\pm\hat{y}$, and $\pm\hat{z}$ directions, as shown in Fig. 9.7, the net effect is a damping force, which slows all the atoms in the gas and reduces its temperature. Since the cooling is dependent on the Doppler effect, the cooling procedure is also called as Doppler cooling. It can be shown that the atoms experience a viscous damping force $F = -\alpha v$ where α is a damping coefficient that is in turn determined by the intensity and frequency of the laser beams, and the lifetime of the excited level. The atomic motion in the field of the laser beams is just like the motion of particles in a viscous fluid. Consequently the system of cooled atoms confined by six laser beams is termed as 'optical molasses'. We may note here that the laser beams used for cooling must have extremely narrow linewidths, typically less than the natural linewidth ($\gamma = \tau^{-1}$) of the atomic transition.

The optical molasses configuration can be easily converted into a trap by adding a spherical magnetic field generated by two identical coils carrying opposite currents. This trap is called as a magneto-optical trap (Metcalf and van der Straten 1999). This is a forerunner of the optical traps and is a workhorse in a laser-cooling laboratory. Typically the size of the cold atoms trapped at the centre of the trap is a fraction of a millimetre, the density is in the range of 10^{11} atoms/cc and temperature of a few tens to hundreds of μK. Fig. 9.8 shows a view of the laser cooling laboratory of BARC where atomic vapours of rubidium and caesium are cooled to temperatures in μK range.

Limits of Atomic Cooling and 'Hybrid' Approach

Laser cooling process has intrinsic cooling limits (Chu 1998; Cohen-Tannoudji 1998; Phillips 1998). The first limit is called, T_D—the 'Doppler cooling limit'. This limit signifies the randomness associated with the photon-absorption process and is given by $k_B T_D = \hbar\gamma/2$. For example, for sodium atoms $T_D = 240\,\mu K$ whereas for cesium $T_D = 120\,\mu K$. The optical molasses under suitable conditions provide an additional mechanism called 'Sisyphus cooling' (Chu 1998) which helps to cool the atoms below T_D and reach the second limit of cooling called 'recoil limit' denoted by T_R. This limit is intrinsic with a laser cooling process and is a measure of the uncertainty in the energy of the atoms caused by the recoil energy (E_R), that is, $k_B T_R = E_R$. As an example the recoil limit for sodium atom is 2.4 μK. Laser cooling technique can be conveniently employed for cooling a vapour or a gas of neutral atoms starting from the room temperature to T_R. Crossing this limit, however, is difficult since the laser cooling process uses momentum exchange between photon and atoms to reduce the kinetic energy of the atoms and each elementary momentum transfer results in recoil energy.

The strategy followed for attaining the ultra-low temperatures required for crossing the phase boundary of BEC is a combination of laser cooling and evaporative cooling and is called as the 'hybrid' approach. In this

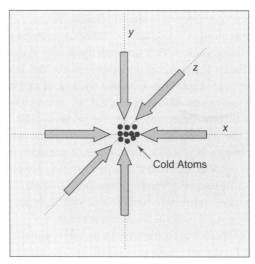

Fig. 9.7 Laser cooling of an isotropic gas using six-beam optical molasses geometry.

Fig. 9.8 Laser cooling facility developed at BARC. The inset shows the fluorescence image of cold atoms at the centre of the trap.

methodology the atoms are first cooled to few tens of μK using laser-cooling technique, specifically a magneto-optical trap and are then transferred into a suitable magnetic trap followed by evaporative cooling to cool below T_R. This strategy is used in almost all experiments aimed at formation of BEC (Anderson et al. 1995; Davis et al. 1995; Bradley et al. 1995). Fig. 9.9 provides a simple illustration of the atomic cooling process using the hybrid approach. The final limit for the temperature of an atomic gas is imposed by uncertainty relation. For example, an atom cooled to $1\ pK (10^{-12}\ K)$ the uncertainty in the velocity is $\sim 3\ \mu$m/s which corresponds to ~ 0.5 cm uncertainty in its position. Spatial delocalization of this magnitude is large compared to the typical sizes of the trapped atomic samples.

BEC IN ATOM TRAPS

An atomic gas of alkali atoms confined in traps represents an inhomogeneous Bose gas of weakly interacting particles. The gas is not homogeneous since it is confined by a spatially varying potential and the inter-particle interaction, though weak, can have marked effect on BEC. Consequently the BEC of atomic gases in atom traps shows many special features, which are often different from those discussed in the standard text books.

PHASE TRANSITION

For an ideal Bose gas of N atoms confined in a harmonic potential the critical temperature is given by (Bagnato, Pritchard, and Kleppner 1987)

$$(T_c)_{\text{Trap}} = (\hbar\omega/k_B)(N/1.202)^{1/3} \qquad (7)$$

where $\omega = (\omega_x \omega_y \omega_z)^{1/3}$ is the geometric mean of the harmonic trap frequencies. This critical temperature is different from that in eq. (2) for a homogeneous gas and is about two times larger than that for a rigid box having the same effective volume as the trap. In fact, a strong confining

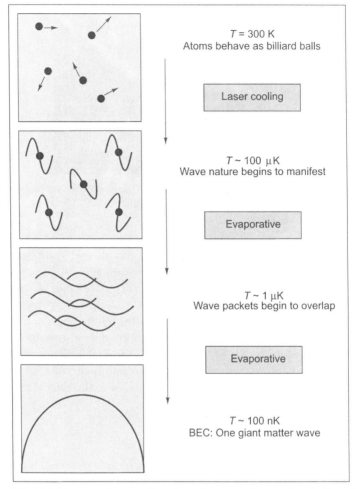

Fig. 9.9 Hybrid approach for cooling atoms from room temperature to the critical temperature required for Bose-Einstein condensation.

potential can allow BEC to occur at a high critical temperature. Further below T_c, the condensate fraction in atom traps varies as

$$\frac{N}{N_0} = 1 - \left(\frac{T}{T_c}\right)^\sigma \quad (8)$$

where σ is related to the 'confining power' of the potential. For a harmonic trap $\sigma = 3$ and that makes eq. (8) different from eq. (4) which applies to a homogeneous gas. Further for a Bose gas confined in a trapping potential, BEC can occur at finite temperature in lower dimensions, which is not the case for a homogeneous gas.

BEC is a condensation phenomenon in which all the atoms collapse into the ground state of the potential confining the gas. In a homogeneous gas the ground state has the same spatial extension as the excited states and as a consequence BEC is a condensation in the momentum space. However, in a harmonic potential the ground state has the smallest spatial dimension and BEC can be observed in configuration space also. In traps the onset of BEC can be identified by observing the collapse of a cloud of atoms into a smaller and denser core surrounded by diffused un-condensed cloud (Anderson et al. 1995; Davis et al. 1995; Bradley et al. 1995). The dense core corresponds to the Bose condensate phase.

Stability of Bose Condensates and Bosnova

The stability of Bose condensates depends on whether the inter-atomic interaction is repulsive or attractive (Griffin, Snoke, and Stringari 1995; Liftshitz and Pitaevskii 1980; Sackett and Hulet 2001). As discussed earlier, the inter-atomic interaction is conveniently expressed in terms of the scattering length a. The sign a is important since it provides information on whether the inter-atomic interaction is repulsive ($a > 0$) or attractive ($a < 0$). In general, for repulsive interaction Bose condensates are stable. The situation, however, is quite different, for a Bose gas with attractive interaction is quite different. Such a gas is not mechanically stable against collapse and BEC may be pre-empted by the first order gas-liquid or gas-solid phase transitions. If, however, the condensate is confined by a potential and the number of atoms forming the condensate (N_0) is not too large, it is possible for the zero point energy to exceed the attractive interaction and stabilize the Bose condensate against the collapse.

The influence of the nature of the interaction, attractive or repulsive, on the stability of the Bose condensate may be understood qualitatively by referring to Fig. 9.10. When the inter-atomic interaction is repulsive, the atoms try to push apart and the condensate expands to a larger volume. The density profile becomes much flatter at the central region for large N_0 and the condensate becomes stable. Examples of such systems are the dilute vapours of ^{87}Rb and ^{23}Na, which have $a > 0$ and as a consequence these Bose systems form stable condensates (Anderson et al. 1995; Davis et al. 1995). When the inter-atomic interaction is attractive, the condensate atoms try to draw together, and thus compressing the condensate to a smaller volume. However, in the ground state of the trap, as the volume of the condensate decreases, the uncertainty principle requires the momentum or velocity to increase. The zero point energy associated with the ground state does not allow the volume to be reduced beyond a certain minimum. This exerts an outward pressure—the Heisenberg pressure, which opposes the compressive force of attraction. For small condensates these two forces are balanced at a particular critical condensate size $(N_0)_c$ and an equilibrium state is reached. However, for $N_0 > (N_0)_c$ the condensate collapses. A gas of ^{7}Li, which has $a < 0$, exhibits behaviour of this kind. In this case $(N_0)_c \sim 1400$ and beyond this the zero point energy is unable to stabilize the condensate against collapse (Bradley et al. 1995; Sackett and Hulet 2001).

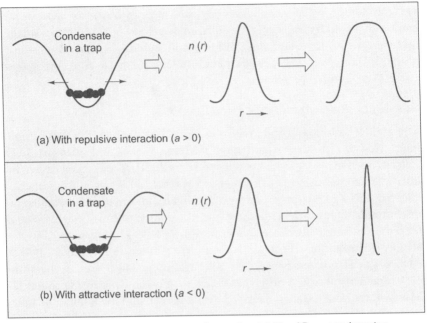

Fig. 9.10 Effect of inter-atomic interaction on the stability of Bose condensates.

Importance of a in BEC is demonstrated by a beautiful experiment done at JILA (Bagneto, Pritchard, and Kleppner 1987). In this experiment a was varied by magnetically tuning a Feshbach resonance of ^{85}Rb atoms. When the magnetic field was 166 G, a became positive and a large and stable condensate was formed in a trap. When the magnetic field was increased to 166.8, a became negative thereby making the condensate unstable. The condensate collapsed in sudden and violent implosion, in which a large fraction of the condensate was blown away with sufficient energy to escape from the trap and a small fraction of the condensate remained at the core. This phenomenon has been named as 'Bosenova' in analogy with the supernova explosions of the stellar systems. It is interesting to note that just before the occurrence of the supernova it is the Pauli pressure, due to exclusion principle, which prevents the collapse of a neutron star. In the case of Bosenova, it is the Heisenberg pressure, due to uncertainty principle, which prevents the collapse of the condensate.

ATOMS AS WAVES

The wave particle dualism as propounded by de Broglie is the basic foundation of quantum physics. Every particle is associated with a de Broglie wavelength Λ, but at normal temperatures the magnitude of Λ is too small to manifest the wave nature, for example, at 100 K, $\Lambda \sim 0.1$ nm, which is comparable to the dimension of the atom itself. When the gas is cooled to ultra-low temperatures Λ becomes macroscopic. For example, at $T = 1\mu$K, $\Lambda = 1\mu$m while at $T = 10$ nK, $\Lambda = 10$ μm. Atomic matter waves of such dimensions can be directly experienced in the laboratory. The wavelike properties of atoms become very dramatic through the phenomenon of BEC, where the atoms become one giant matter wave, all oscillating in phase (Ketterle 2002, Nobel lecture). This coherent matter wave behaves entirely like a laser and exhibits all phenomena that fall in the domain of 'photon' optics, such as interference, diffraction, amplification, etc. This new field of research is coherent 'atom optics'—optics with coherent matter waves. Interestingly, the notion of wave and matter gets interchanged in going from photon optics to atom optics. In photon optics the electromagnetic waves are manipulated using optical elements that are made up of matter, for example prisms, lenses, beam splitters, etc. In case of atom optics, the analogous optical elements for manipulation of matter waves are provided by the electromagnetic fields. Few years ago, the idea that atoms can exhibit laser like character was somewhat unthinkable. Today, BEC has made this idea a reality.

Coherence

The coherence property of Bose condensates has been a subject of intense theoretical discussions (Griffin, Snoke, and Stringari 1995). It is related to the nature of the phase of the Bose condensate and to the 'spontaneous symmetry breaking' in mesoscopic systems. An important question is whether the phase is merely a mathematical construction or it can be actually observed in an experiment. Measurement of an absolute phase is possible only if there exists a phase reference. This implies that only a relative phase can be established by performing an interference experiment wherein a pair of statistically independent and physically separated condensates is allowed to overlap to produce interference fringes. It is a non-trivial question whether two independent condensates have a well-defined relative phase. Even if the condensate has no phase information initially, a relative phase may be established as a result of the experimental process. This is an example of 'spontaneous symmetry breaking'. Every experimental realization will show a distinct relative phase, however, this phase is random so that the symmetry of the system is broken in a single experiment. It is, however, restored if an ensemble is considered. This implies that in case of two independent overlapping condensates, a high contrast interference pattern can be seen, but the phase of the pattern should vary randomly between different experiments.

The evidence for coherence in Bose condensates was provided by an interference experiment at MIT (Andews et al. 1997), where high contrast interference between two condensates was observed. A schematic representation of this experiment is shown in Fig. 9.11. In this work two independent condensates of ^{23}Na were produced in a double wall potential that is constructed by partitioning a magnetic trap by a sheet of blue detuned laser to provide repulsive dipole force. After the trap was switched off the condensates fell down due to gravity, expanded ballistically, and eventually overlapped. The interference pattern was observed using a resonant light beam, which was absorbed strongly at the interference maxima. This experiment provided a direct evidence for first order coherence and existence of a relative phase of two condensates.

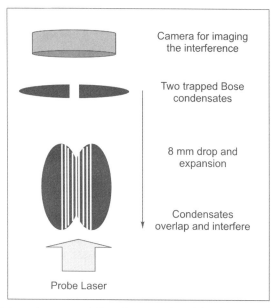

Fig. 9.11   Schematic representation of an experiment for demonstration of the first order coherence of Bose condensates.

In optics the first order coherence function is not enough to distinguish between a coherent laser field and a filtered chaotic light from a thermal source since both give rise to the interference pattern in Young's double split experiment. We therefore require the second order coherence function, $g^{(2)}(\tau)$, which describes the correlation between two separate photon detection events separated by time τ and provides a clear contrast between quantum mechanical and classical description of the radiation field. In an analogous manner, in case of atoms, the second order correlation function is expected to show the distinct quantum properties of BEC vs those of matter waves from a thermal source. The mean field energy of the condensate provides a direct measure of $g^{(2)}(\tau)$. These experiments show that $g^{(2)}(0) = 1$ for the condensate which is different from $g^{(2)}(0) = 2$ for the thermal gas (Ketterle and Miesner 1987). The third order coherence function depends on the probability of finding three atoms near each other and as a consequence governs the rate of three body recombination rate. Theoretical calculations show that the three-body recombination rate in Bose condensates is three times smaller than that for a thermal cloud at the same mean density. These predictions have been experimentally validated in an experiment at JILA (Burt et al. 1997). The existence of the higher order coherence implies that the density fluctuations in the Bose condensate are suppressed relative to those in a thermal gas with the same mean density. The experimental observations accumulated so far are consistent with the standard assumption that a Bose condensate is coherent to first and higher order, and can be characterized by a macroscopic wavefunction.

Stimulated Scattering

The process of stimulated emission of photons is known from the theory of lasers. This process provides a gain mechanism in an optical laser, that is, the presence of photons in the lasing mode stimulates the emission of more photons into it. Stimulated emission of photons is a special case of Bose stimulation, which occurs for scattering processes involving Bosons. This is a fallout of the Bose statistics, which implies that the transition rate in a state having N bosons is proportional to $(N + 1)$ and the occupied state 'attracts' other bosons to the same state. Bose stimulation in atoms has been reported by MIT group (Miesner et al. 1998; Chikkatar et al. 2000) by observing the growth of Bose condensate in presence of thermal atoms. The experimental observations fit very well with the theoretical result that the Bose stimulation rate is proportional to $(N + 1)$ where N is the number of atoms already present in the condensate. Just as the stimulated emission provides a gain mechanism for photon lasers, Bose stimulation in a Bose gas constitutes the gain mechanism for an 'atom laser'.

In Bose condensates, it is possible to form molecules coherently through stimulated recombination in a two-body collision process, that is, $Rb + Rb \to Rb_2$. In such a situation the atoms and recombined molecules form a two-phase Bose condensate (Wynar et al. 2000). Since the molecular condensate represents a coherent molecular matter wave, this process is the counterpart of the optical frequency doubling in the matter wave domain.

Atom Laser

Atom laser refers to the coherent matter waves of bosonic atoms just as the laser is a coherent beam of photons. The interference phenomenon between Bose condensates establishes the coherent nature of the matter in BEC and leads to the idea of 'atom laser', that is, coherent atomic beam generators. A practical realization of such a device, however, requires development of output coupling of coherent atoms from the condensates. In a photon laser, the laser cavity formed by two or more high reflecting mirrors confines the photons. The output coupling is provided by a partially transmitting cavity mirror, which leaks out a small part of the coherent cavity radiation as an output of the device. The magnetic trap that is used for confining the Bose condensed atoms plays the same role as optical cavity for photons. To form an atom laser one needs to devise an appropriate output-coupling scheme by which a part of the condensate can be leaked out of the trap.

In the first experimental demonstration of an atom laser (Mewes et al. 1997), the MIT group developed a scheme of output coupling in sodium BEC by exposing the Bose condensate to *RF* radiation as shown schematically in Fig. 9.12. The role of the *RF* is to induce magnetic transition between Zeeman states of the atom in such a way that the atoms make transition from the 'trapped' state to an 'un-trapped' state. The fraction of the condensate that is converted in the un-trapped state is no longer confined by the magnetic trap and drops out of the trap. This forms the coherent matter output of the atom laser. The output of the first atom laser had a relatively large divergence due to the inter-atomic repulsion, which pushes the atoms apart as soon as they are released from the trap. This beam divergence problem was overcome at National Institute of Standards and Technology (NIST) by introducing an ingenious output-coupling scheme (Hagley et al. 1999). In this scheme the transition from the 'trapped' state to an 'un-trapped' state is induced by stimulated Raman process instead of direct *RF* excitation. The scheme is illustrated in Fig. 9.13.

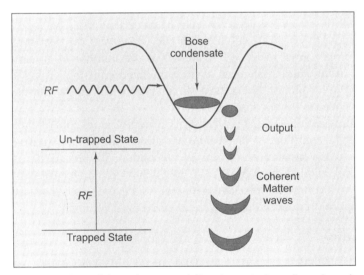

Fig. 9.12   Atom laser. Schematic representation of an output coupling scheme based on *RF* excitation.

It employs two counter-propagating laser beams having linear and mutually orthogonal polarizations, slightly different frequencies, and passing through a trapped Bose condensate of sodium atoms. The difference $\Delta \nu$ is matched exactly with the energy difference between the trapped and un-trapped states. The atoms in the condensate undergo Raman transition from the trapped state to the un-trapped state by exchanging photons between the two lasers, that is, absorption from one laser and stimulated emission into the

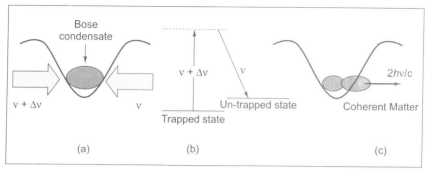

Fig. 9.13 Atom laser. Output coupling scheme based on Raman transition. (a) The trapped Bose condensate interacts with two counter propagating lasers. (b) The lasers cause Raman transition to convert a part of the condensate to an un-trapped state. (c) The un-trapped part of the condensate is released with a momentum of two photons.

other laser. As a result the atoms not only become free from the trap but they also acquire momentum $\sim 2h\nu/c$ in the forward direction. Because of this relatively large forward momentum, the atoms travel a sufficiently long distance before spreading in the transverse direction by the inter-atomic repulsive forces. This results in a divergence of the output as small as a few milliradians, which is comparable to the divergence of a conventional photon laser.

There are several similarities between an atom laser and a photon laser, and these are summarized in Fig. 9.14. At the same time, the atom laser and photon laser have several distinguishing features. First, unlike the photons in a laser cavity, the atoms in the Bose condensate interact with each other. Second, the atoms are massive and therefore a beam of atom laser is influenced by gravity. This aspect becomes interesting in the context of atom interferometry and accurate measurement of gravity. Third, atom lasers are always single mode lasers because Bose condensates occupy the lowest energy level of the trapping potential well. In contrast, the photon lasers operate in very high order modes and also multimode operation is quite common. Finally, BEC takes place in the conditions of thermal equilibrium. The photon laser, on the other hand, always operates in non-equilibrium conditions due to the requirement of population inversion.

Non-linear Atom Optics

An atom laser can be used to investigate non-linear phenomenon involving matter waves. This has been demonstrated at NIST by performing four-wave mixing of matter waves in analogy with the optical four-wave mixing (Deng et al. 1999). The origin of the non-linearity required for four-wave mixing of matter waves is in the Gross-Pitaevskii (GP) equation that describes the space and time evolution of the wavefunction $\psi(r,t)$ of the Bose condensate. The

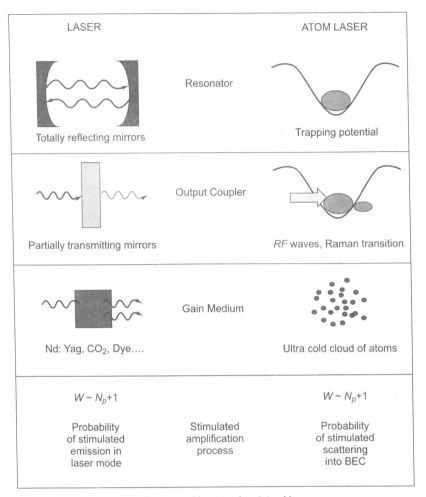

Fig. 9.14 A comparison of a 'photon' laser and an 'atom' laser.

GP equation for a Bose condensate in a trap is (Lifshitz and Pitaevskii 1980)

$$i\hbar \frac{\partial \psi(r,t)}{\partial t} = \left[-\frac{\hbar^2}{2M}\nabla^2 + V_{\text{trap}} + NU\,|\psi(r,t)|^2 \right] \psi(r,t) \qquad (9)$$

where V_{trap} is the trapping potential and the inter-atomic interaction is represented by a mean-field potential which is proportional to the number of atoms in the condensates and the interaction strength $U = 4\pi\hbar^2 a/M$. An examination of eq. (9) shows that the last term is similar to the optical third order non-linear term $\chi^{(3)} E^2 E$ which is responsible for the four-wave mixing of light waves.

Fig. 9.15 shows a schematic representation of four-wave mixing experiment with matter waves. A Bose condensate is illuminated by pulses of two laser beams that drive stimulated Raman process to release a matter

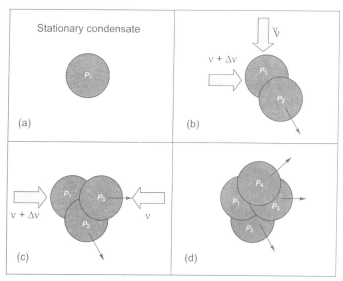

Fig. 9.15 Degenerate four-wave mixing of matter waves. (a) to (c): From a stationary condensate in zero momentum state ($P_1 = 0$) two components with different momenta P_2 and P_3 are generated by Raman transitions caused by two laser beams of frequencies ν and $\nu + \Delta\nu$. (d) Three components with momenta P_1, P_2, and P_3 interact and produce a fourth component with momentum P_4.

wave output. Initially the lasers are arranged perpendicular to each other. Atoms of the condensates acquire a momentum of two photons by absorption from one pulse and stimulated emission from the second. Thus a part of the condensate is set into motion. At this stage the laser configuration is changed from perpendicular to counter propagating to set apart another part of the condensate into the motion with a different direction and momentum. The remaining stationary part of the condensate interacts with these two moving parts through the non-linear interaction term of the GP equation to generate the fourth part of the condensate. This is similar to the interaction of three optical waves through the non-linear interaction term to generate a fourth wave in the four-wave mixing process. The only difference is that while the optical waves need a non-linear medium for the four-wave mixing process to take place, no such medium is required for four-wave mixing of matter waves.

The structure of the GP equation is such that it also supports solitons, that is stable localized waves, which propagate in a non-linear medium without spreading. For condensates with positive scattering length the solitons are 'dark solitons', which represent a 'notch' with a characteristic phase step. Such solitons have been demonstrated at NIST (Denschlag et al. 2000).

Atom Amplification

Atom amplification is a process similar to light amplification in a photon laser. In optical domain, amplification is realized by use of active optical elements such as pulsed-dye amplifiers, which consist of a gain medium for amplification process. By passing through such a device a weak laser beam can be amplified while maintaining its phase coherence. If we carry this analogy to the atomic domain then an 'atom amplifier' is a device that converts atoms from the active medium into an atomic matter wave that is exactly in the same quantum state as the input wave. Amplifying atoms is more subtle than amplification of photons since the atoms cannot be created or destroyed like photons, they can only change their quantum states.

An atom amplifier requires an active medium of ultra-cold atoms that have very narrow spread of velocities and Bose condensate is a natural choice. Further one must provide a coupling mechanism that transfers atoms from the active medium to an input mode while conserving energy and momentum. Scattering of laser light provides this transfer of atoms. The recoil associated with this scattering process accelerates some atoms to exactly match the velocity of the input atoms—the process referred to as the super-radiance in Bose condensates. This results in the amplification of the atoms and further the amplified atoms acquire the same quantum mechanical phase as that of the input atoms (Stenger et al. 1999; Stamper Kurn et al. 1999; Inouye et al. 1999; Ovchinnikov et al. 1999; Kozuma et al. 1999). This basic scheme of atom amplification is illustrated in Fig. 9.16.

We may now quickly look at the super-radiance process that arises due to scattering of photons by Bose condensate. When a non-spherical Bose condensate is exposed to a laser light it results in separation of a part of the condensate with well defined velocity. This is the phenomenon of super-radiance in Bose condensates and is analogous to the optical parametric amplification in the optical domain. Qualitatively the super-radiance phenomenon may be understood as follows: When a laser is incident on a Bose condensate, it causes excitation followed by spontaneous emission that result in imparting recoil momentum to some atoms in the condensates. The matter wave corresponding to these higher momentum states interfere with the matter waves of the rest of the condensate to form a matter wave grating. This grating diffracts more laser photons and the change in the photon momentum during diffraction is transferred in turn to the matter wave grating. This results in scattering of more atoms in the higher momentum state. Since the phase of the scattered photons is preserved in the scattering process, the population in the higher momentum state is phase coherent. However, since the initial scattering of the atoms is random due to random nature of the spontaneous emission process, atoms get scattered into various random momentum states. In a spherically symmetric condensate, there is no preferred direction and the randomness in the occupation of various momentum states continues. In magnetic traps employed for BEC, however,

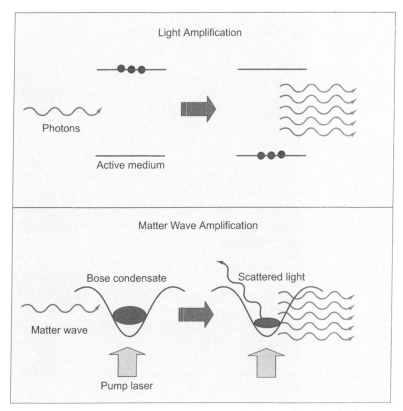

Fig. 9.16 Amplification of matter waves compared to the amplification of light.

the Bose condensate is elongated along the axis of the trap. In such a condensate the growth rate of the atoms with momentum along this axis is higher. Atoms are then preferentially scattered into these momentum states, which eventually dominate over other states. In other words the gain available for the growth of matter waves with momentum along the axial direction is more compared to the gain available in other directions. The matter wave corresponding to these momentum states is the super-radiant matter wave, which separates out from the rest of the condensate. The situation is exactly similar to the light emission from an optical gain medium in a laser whose physical dimensions in the axial direction are more compared to that in the transverse direction.

The atom amplification process does not destroy the coherence. This has been verified by using a matter wave Mach-Zehnder interferometer for measurement of the phase of the amplified matter wave. Functionally a matter wave Mach-Zehnder interferometer is an exact analogue of optical Mach-Zehnder interferometer; it splits the matter wave into two parts, allows them to travel through two different paths, combines them again and measures the relative phase (Stenger et al. 1999; Stamper-Kurn et al. 1995; Inouye et al. 1999; Ovchinnikov et al. 1999; Kozuma et al. 1999; Kozuma et al. 1999).

FUTURE TECHNOLOGIES

The field of ultra-cold atoms including BEC is no longer a subject of academic interest alone. In very recent years the focus is shifting to applications of ultra-cold atoms and Bose condensates for development of newer devices and technologies. An atomic clock based on the microwave transition in the ground hyperfine states of cesium atoms is perhaps the first application of ultra-cold atoms. This clock measures frequency with an accuracy of 2 to 3 parts in 10^{14} which corresponds to time accuracy of 2 ns per day or one second in 1,400,000 years. In recent years the ultra-cold atoms are providing a platform for many quantum technologies that include quantum computer, nanofabrication, gravity gradiometers, etc.

Quantum Computer with Cold Atoms

Past several years, we have seen an enormous growth in computing power and information processing speed. This is summed up in the familiar Moore's law—'Number of transistors on a chip double every two years' as stated by Moore in 1965. The question that is often asked is 'Can this doubling continue with ever increasing functionality?' If the current growth continues then a 'bit' will shrink to the size of an individual atom/molecule in a decade from now. At these atomic scales the laws of quantum mechanics dominate and the notion of classical bit does not hold good. Here lies the concept of quantum computer (Nelsen and Chuang 2002), which focuses on the pure quantum features, that is superposition and entanglement, of a collection of atom-sized bits—the quantum bits or the 'qubits'. The hardware for such a machine is given by the individual laser-cooled atoms and isolated photons. Fig. 9.17 illustrates the concept of a qubit for a collection of ultra-cold atoms. For an N qubit the most general state is defined by 2^N independent amplitudes, which in a sense represent the quantum of stored information. Consider now a quantum computer that is configured with 500 ultra-cold atoms. The number of independent amplitudes in this case are $2^{500} \sim 10^{150}$ which is much larger than the number of silicon atoms in the universe. This implies that a quantum computer based on 500 ultra-cold atoms will have far superior information storage capability compared to a classical computer.

A proposed configuration of a quantum computer using ultra-cold atoms (Monroe 2002) is described in Fig. 9.18. The concept is based on 'optical lattice', which consists of periodic light-shift potentials. These wavelength sized micro potentials are formed by the interference of two or more polarized laser beams. Atoms are cooled and localized in the potential minima. The micro-potentials being shallow, the atoms need to be cooled to ultra-low temperatures. The shape of an optical lattice can be easily tailored by varying the parameters of the laser beams and also by applying a magnetic field.

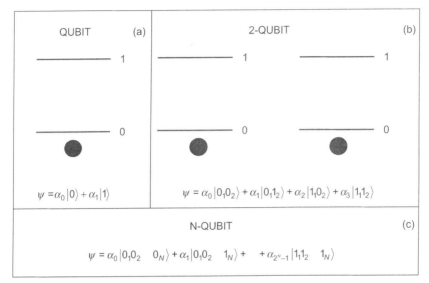

Fig. 9.17 Quantum Bit. States 0 and 1 of an atom are taken as the fundamental concept of information—(a) In case of a single atom the wavefunction ψ is a superposition of states 0 and 1, and α_0 and α_1 are the corresponding amplitudes. (b) For two atoms the superposition consists of four states denoted by $|A_i\ B_j\rangle$ where $A, B = 0, 1$ is the state index and $i, j = 1, 2$ is the atom index. For example the state $|0_1\ 1_2\rangle$ indicates that the atom-1 is in state $|0\rangle$ and the atom-2 is in state $|1\rangle$. The total number of amplitudes is 4. (c) A general N-qubit consists of 2^N amplitudes.

Optical lattice exhibits many features that are associated with solid-state lattice. The important difference, however, is that the optical lattice has a long coherence time.

Construction of a quantum computer requires (1) a scalable physical system with well-characterized qubits; (2) an ability to initialize the state of the qubit to a simple fiducial state; (3) relatively long decoherence times much longer than the gate operation time; (4) a universal set of quantum gates; (5) a qubit-specific measurement capability; (6) ability to inter-convert stationary and flying qubits (photons); and (7) ability to faithfully transmit flying qubits between specific locations. In many of these regards a quantum computer based on neutral atoms trapped in an optical lattice scores over the other proposals, which include, for example, nuclear magnetic resonance or cavity quantum electrodynamics based quantum computation. In addition, quantum computer based on neutral atoms can offer scalability and massive parallelism due to the optical lattice geometry.

Nano-fabrication with Cold Atoms

Taken to its absolute limit, nano-technology involves placing single atoms one by one in desired locations. This is possible by the technique of 'atom

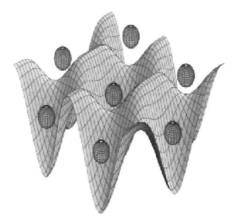

Fig. 9.18 Schematic representation of an optical lattice loaded with atoms.

lithography' that is based on deposition of atoms cooled and manipulated by laser light. For example, neutral atom traps can be used to controllably capture single atoms at a time and then transport them to a desired location, or a beam of cold neutral atoms can be spatially segregated through interaction with an electromagnetic stationary wave and deposited onto a substrate, or a matter wave can be manipulated using atom optics to generate desired patterns (McClelland 2000; Meschede and Metcalf 2003). In recent years, concerted efforts are being made to use atom lithography to develop nano-fabrication technologies in < 10 nm range, that is interferometric precision, for production of unique devices relevant in electronic, opto-electronics, and information technology. These include nano-sized photon detectors/emitters, nano-structure photonic crystals, and nano-islands.

'Atom chip' provides a new outlook for neutral atom manipulation using integrated micro-devices (Folman et al. 2000). In atom chip, the atoms are cooled and trapped with micron-sized wires fabricated on a highly reflecting surface. This device can confine atoms at a fixed distance from the surface, typically a few microns, therefore the name 'atom chip'. They have several advantages—low power consumption, light confinement, and precision micro-scale control. These atom chips are similar to the integrated circuits in electronics and form basis for applications of cold atoms ranging from atom optic to quantum information processing.

Atom Interferometry

Atom interferometry is a novel and emerging quantum technology that has the potential to provide new gravity measurement capability (Mlynek et al. 1992). The measurement of gravity and its gradient has many important scientific and technological applications. These include for example, measurement of gravitational constant G, test of general theory of relativity, underground structure determination, oil and mineral exploration, and geodesy. Atom interferometry is based on the wave-particle duality. The wave like nature of the ultra-cold atoms is exploited to construct an atom interferometer and the particle-like character makes the atom wave interference sensitive to gravity, thus providing a way to measure the gravity and its gradient. The great advantage of atom interferometry can be appreciated by the fact that it has an inherent inertial sensing sensitivity that is more than 10 orders of magnitude greater than an equivalent laser interferometer. A technical problem associated with the characterization of gravitational forces is rooted in the equivalence principle—an inability to distinguish gravitationally induced acceleration from the accelerations of the reference frame of the measurement. This difficulty is overcome with gradiometric measurements in which two simultaneous spatially separated acceleration measurements are made with respect to a common reference platform. Such measurements have already reported differential acceleration accuracy of $< 10^{-9}$g. It is expected that an accuracy of 10^{-13}g is realizable in near future.

CONCLUSION

BEC has come a long way from being merely a theoretical curiosity. At least on one count Einstein was wrong—that he had some doubts about the phenomenon that he himself had discovered. Today BEC is not only a reality but it has become the laboratory for many body physics (Anglin and Ketterle 2002; Sackett and Hulet 2001) and coherent atom optics (Rolston and Phillips 2002), and further entering in the molecular domain (Bernett et al. 2002). BEC in alkali vapours has already gone beyond the confirmation of old theories and has provided enough fodder for newer theoretical developments. The field is expanding in several directions including quantum degeneracy for the fermionic systems, vortices, superfluidity, Josephson tunnelling and quantum phase transitions (Anglin and Ketterle 2002). BEC has continued to change our notions about matter and coherent matter waves. It is now possible to develop non-linear and quantum atom optics which are de Broglie wave analogue of nonlinear and quantum optics (Rolston and Phillips 2002). These include phenomena such as four-wave mixing, amplification, soliton propagation, and squeezing of coherent matter waves. At the same time the entire technology of ultra-cold atoms developed for BEC is now driving many frontier areas of research including non-accelerator particle physics, quantum entanglement, Bell's inequality, and quantum teleportation, quantum information processing, nanotechnology, ultra-cold chemistry, etc., in a manner that the BEC and ultra-cold atoms are certain to dominate the basic research in physics in coming years.

BEC has provided a new vision for basic research and technology development in very recent times and all this has originated from the six months of work that Einstein did in 1924. Abraham Pais in his work on Einstein's life and science describes this as 'Such is the scope of his oeuvre that his discoveries in those six months do not even rank among his five main contributions, yet they alone would have sufficed for Einstein to be remembered for ever (1982). The best possible way to end this chapter is to quote Einstein himself—Einstein while receiving the second Planck medal in 1929 said, 'I am ashamed to receive such a high honour since what all I have contributed to quantum physics are occasional insights arose in the course of fruitless struggle with the main problem'. An example of such occasional insights of Einstein is the phenomenon of BEC, which lured researchers since 1925 and will continue to keep them occupied for many years to come.

REFERENCES

Anderson, M.H., J.R. Ensher, M.R. Mathews, C.E. Wieman, and E.A. Cornell (1995). *Science*, 269, 198.

Andews, M.R., C.G. Townsend, H.J. Miesner, D.S. Durfee, D.M. Kurn, and W. Ketterle (1997). *Science*, 275, 634.

Anglin, J.R. and W. Ketterle (2002). *Nature*, 416, 211.

Bagnato, V., D.E. Pritchard, and D. Kleppner (1987). *Phys. Rev.*, A 35, 4354.

Bernett, K., P.S. Julienne, P.D. Lett, E. Tiessinga, and C.J. Williams (2002). *Nature*, 416, 225.

Bose, S.N. (1924). *Z. Phys.*, 26, 178.

Bradley, C.C., C.A. Sackett, R.J. Tollet, and R.G. Hulet (1995). *Phys. Rev. Lett.*, 75, 1687.

Burt, E.A., R.W. Ghrist, C.J. Myatt, M.J. Holland, E.A. Cornell, and C.E. Wieman (1997). *Phys. Rev. Lett.*, 79, 337.

Chikkatur, A.P., A. Gorlitz, D.M. Stamper-Kurn, S. Inouye, S. Gupta, and W. Kettrele (2000). *Phys. Rev. Lett.*, 85, 483.

Chu, S. (1998). *Rev. Mod. Phys.*, 70, 685.

Cohen-Tannoudji, C. (1998). *Rev. Mod. Phys.*, 70, 707.

Cornish, J.L., N.R. Claussen, J.L. Robert, E.A. Cornell, and C.E. Wieman (2000). *Phys. Rev. Lett.*, 85, 1795.

Davis, K.B., M.O. Mewes, M.R. Andrews, N.J. van Druten, D.S. Durfee, D.M. Kurn, and W. Ketterle (1995). *Phys. Rev. Lett.*, 75, 3969.

Deng, L., E.W. Hagley, J. Wen, M. Trippenbach, Y. Band, P.S. Julienne, J.E. Simsarian, K. Helmerson, S.L. Rolston, and W.D. Phillips (1999). *Nature*, 398, 218.

Denschlag, J., J.E. Simsarian, D.L. Feder, C.W. Clark, L.A. Collins, J. Cubizolles, L. Deng, E.W. Hagley, K. Helmerson, W.P. Reinhardt, S.L. Rolston, B.I. Schneider, and W.D. Phillips (2000). *Science*, 287, 97.

Einstein, A. (1924). *Sitz. Preuss. Akad. Der Wiss*, 261.

———— (1925). *Sitz. Preuss. Akad. Der Wiss*, 3.

Folman R., P. Kruger, D. Cassettari, B. Hessmo, T. Maier, and J. Schmiedmayer (2000). *Phys. Rev. Lett.*, 84, 4749.

Griffin, A. (1993). *Excitations in a Bose Condensed Liquid*. Cambridge: Cambridge University Press.

Griffin, A., D.W. Snoke, and S. Stringari (1995). *Bose Einstein Condensation*. Cambridge: Cambridge University Press.

Gyeytak T.J. and D. Kleppner (1984). In G. Grynberg and R. Stora, (eds), *New Trends in Atomic Physics*, 2, 1127, North Holland, Amsterdam.

Hansch, T. and A. Schawlow (1975). *Opt. Comm.*, 13, 68.

Hess, H.I. (1986). *Phys. Rev.*, B 34, 3476.

Hagley, E.W., L. Deng, M. Kozuma, J. Wen, K. Helmerson, S.L. Rolston, and W.D. Phillips (1999). *Science*, 283, 1706.

Inouye, S., A.P. Chikkatur, D.M. Stamper-Kurn, J. Stenger, D.E. Pritchard, and W. Ketterle (1999). *Science*, 285, 5427.

Ketterle, W. (2002). *Rev. Mod. Phys.*, 74, 1131.

Ketterle, W. and H.J. Miesner (1987). *Phys. Rev.*, A 56, 3291.

Kozuma, M., L. Deng, E.W. Hagley, J. Wen, R. Lutwak, K. Helmerson, S.L. Rolston, and W.D. Phillips (1999). *Phys. Rev. Lett.*, 82, 871.

Kozuma, M., Y. Suzuki, Y. Torii, T. Sugiura, T. Kuga, E.W. Hagley, and L. Deng (1999). *Science*, 286, 2309.

Lifshitz, E.M. and L.P. Pitaevskii (1980). *Statistical Physics*, Vol. II, Oxford: Pergamon.

London, F. (1938). *Nature*, 141, 643, *Phys. Rev.*, 54, 1747.

Metcalf, H.J. and P. van der Straten (1999). *Laser Cooling and Trapping*. New York: Springer Verlag

Mewes, M.O., M.R. Andrews, D.M. Kurn, D.S. Durfee, C.G. Townsend, and W. Ketterle (1997). *Phys. Rev.*, Lett. 78, 582.

Miesner, H.J., D.M. Stamper-Kurn, M.R. Andews, D.S. Durfee, S. Inouye, and W. Ketterle (1998). *Science*, 279, 1005.

McClelland, J.J. (2 000). In H.S. Nalwa (ed.), *Handbook of Nanostructure Materials and Nanotechnology*, Vol. I. San Diego: Academic Press.

Meschede, D. and H. Metcalf (2003). *J. Phys. D. Appl. Phys.*, 38, R17.

Mlynek, J., V. Balykin, and P. Meystre (eds) (1992). Special Issue on 'Atom Optics and Atom Interlerometry', *Appl. Phys.*, B. 54, 319-491.

Monroe, C. (2002). *Nature*, 416, 238.

Nielsen, M.A. and I.L. Chuang (2002). *Quantum Computation and Quantum Information*. Cambridge: Cambridge University Press.

Ovchinnikov, Yu B., H.Muller, M.R. Doery, E.J.D. Vredenbregt, K. Helmerson, S.L. Rolston, and W.D. Phillips (1999). *Phys. Rev. Lett.*, 83, 284.

Pais, A. (1982). *'Subtle is the Lord': The Science and the Life of Albert Einstein*. Oxford: Oxford University Press.

Phillips, W.D. (1998). *Rev. Mod. Phys.*, 70, 721.

Rolston, S. L. and W.D. Phillips (2002). *Nature*, 416, 219.

Sackett, C.A. and R.G. Hulet (2001). *J. Opt. B: Quant. Semiclass.* Opt. 3, 1.

Silvera, I.F. and J.T.M. Walraven (1986). 'Spin-polarized Atomic Hydrogen', in D. Brewer (ed.), *Progress in Low Temperature Physics*, 10, 139.

Stenger, J., S. Innouye, A.P. Chikkatur, D.M. Stamper-Kurn, D.E. Pritchard, and W. Ketterle (1999). *Phys. Rev. Lett.*, 82, 4569.

Stamper-Kurn, D.M., A.P. Chikkatur, A. Gorlitz, S. Inouye, S. Gupta, D.E. Pritchard, and W. Ketterle (1999). *Phys. Rev. Lett.*, 83, 2876.

Whitrow, G.J. (1967). *Einstein: The Man and His Achievements*. New York: BBC.

Wynar, R., R.S. Freeland, D.J. Han, C. Tyu, and D.J. Heinzen (2000). *Science*, 287, 1017.

Contributors

ABHAY VASANT ASHTEKAR has served as professor in Universite de Paris VI and Syracuse University of USA. Ashtekar was elected to the Governing Council of the International Society for General Relativity and Gravitation in 1989 for a ten year period. He has been a visiting professor to various Indian and foreign institutes, prominent among them are Ohio University; British Science and Engineering Research Council; Sir C.V. Raman Chair of the Indian Academy of Sciences; The Kramers Chair of Theoretical Physics, University of Utrecht; Guest Scholar at the Max Planck Institut fur Physik und Astrophysik, Munchen, Germany; Visiting Scientist, Raman Research Institute, Bangalore; UICAA, Pune and Golden Jubilee Visiting Professor, Physical Research Laboratory, Ahmedabad. He has given distinguished lectures at the Institute for Fundamental Theory, University of Florida and Center for Theoretical Physics, University of Maryland.

SHASHIKUMAR MADHUSUDAN CHITRE, after his M.A., Ph.D. from Cambridge, served as a Lecturer in Applied Mathematics at the University of Leeds, England during the period 1963–6 and was later a Research Fellow at the California Institute of Technology, Pasadena, USA before joining TIFR in 1967 from where he retired as Senior Professor in 2001. He is currently Distinguished Faculty at the Centre for Basic Sciences at the University of Mumbai. He was a UGC National Lecturer in Physics during 1975–6, and has held visiting positions at Universities of Cambridge, Princeton, Sussex, Amsterdam, Columbia, and Virginia, and was a Max-Planck Fellow at the Max-Planck Institute fur Extraterrestrische Physik, Munich and held Senior Research Associateship of the National Academy of Sciences, USA at Goddard Space Flight Centre, NASA. He is a Fellow of the Royal Astronomical Society, Member of the International Astronomical Union and has been elected to the Fellowships of the Indian Academy of Sciences, the Indian National Science Academy, the National Academy of Sciences, India, the Third World Academy of Sciences and the Maharashtra Academy of Sciences.

NARESH DADHICH has been on the faculty of the IUCAA since its inception in 1988 where he initially served as its project coordinator and was its Director from July 2003 to August 2009. He is a theoretical physicist whose interests lie in the areas of classical and quantum gravity and relativistic astrophysics, particularly their mathematical aspects and has published over 100 papers in international front rank journals in collaboration with many students and colleagues. He has visited several universities and institutes all over the globe. He has been elected to the Council of the International Relativity Society and has been the President of the Indian Relativity Society. He is a deep thinker with a commitment not only to make complex and subtle issues of science accessible to general audience but also to have a dialogue on issues of greater and wider social relevance.

ARVIND KUMAR is DAE Raja Ramanna Fellow at the Homi Bhabha Centre for Science Education (Tata Institute of Fundamental Research, Mumbai). He taught for nearly 12 years at the University Department of Physics, Mumbai. In 1984, he joined the Homi Bhabha Centre for Science Education and was its Director from 1994 till his retirement in 2008. He has published research and supervised Ph.D. theses in the areas of particle theory, quantum black holes, atomic and optical physics, mathematical sociology, and physics education. He played a central role in launching the science Olympiad movement in the country in 1997. He was on the Steering Committee of the National Curriculum Framework-2005.

B.N. JAGATAP is working as Outstanding Scientist in Bhabha Atomic Research Centre (BARC) and is associated with the laser programme of BARC since 1977. He is Senior Professor in Homi Bhabha National Institute and Visiting Distinguished Professor (Adjunct) in the Department of Physics, Indian Institute of Technology Bombay, Mumbai. He worked as visiting scientist at the University of Western Ontario, Canada in 1998 and as the senior visiting fellow at the Centre for Chemical Physics, Canada during 2000–1. He is the recipient of Dr Homi Bhabha Award for Science & Technology (1999). His research interests include laser spectroscopy, quantum optics, ultra-cold atoms, Bose-Einstein condensation and physics of laser elective processes.

JAYANT VISHNU NARLIKAR is Emeritus Professor at the Inter-University Centre for Astronomy and Astrophysics (IUCAA), Pune. He served as the Fellow of King's College at Cambridge (1963–72) and was founder staff member of the Institute of Theoretical Astronomy (1966–72). Narlikar joined TIFR in 1972 where under his charge the Theoretical Astrophysics Group acquired international standing. In 1988 he was invited by the University Grants Commission as Founder Director to set up the proposed IUCAA that under his direction acquired reputation as a worldwide centre for excellence in teaching and research in astronomy and astrophysics. Narlikar is internationally known for his work in cosmology, in championing models alternative to the popularly believed big bang model. He is also known for his efforts for science popularization.

THANU PADMANABHAN, Distinguished Professor and Dean of Core Academic Programmes at IUCAA, Pune, is a world renowned theoretical physicist and cosmologist who has authored over 200 research papers and nine books including six graduate-level textbooks published by Cambridge University Press (CUP). His research work has won prizes from the Gravity Research Foundation (USA) five times including the First Prize in 2008. An elected Fellow of all the three Academies of Science in India, he is currently the President of the Cosmology Commission of the International Astronomical Union and was a Sackler Distinguished Astronomer of the Institute of Astronomy, Cambridge. He has won numerous awards including Shanti Swarup Bhatnagar Award (CSIR), The Millennium Medal (CSIR), G.D. Birla Award, INSA Vainu-Bappu Medal, Al-Khwarizmi International Award, Miegunah Fellowship of University of Melbourne, Infosys Science Foundation prize for Physical Sciences. He received the Padma Shri, the medal of honour from the President of India in recognition of his achievements in 2007.

VIRENDRA SINGH is presently Honorary Professor and INSA Honorary scientist at the Tata Institute of Fundamental Research (TIFR), Mumbai. He was earlier Professor of eminence and Director, INSA C.V. Raman Professor at the same institute .He has held various academic positions in foreign institutions such as the Rockefeller University, the Institute of Advance Study, Princeton, the California Institute of Technology, Pasadena and the Lawrence Radiation Laboratory, Berkeley, USA. Professor Singh is internationally renowned for his outstanding contributions to the development of the theory of elementary particles. His contributions extend over a wide range from the deepest conceptual ideas in physics such as his work on the Foundations of Quantum Mechanics to High Energy phenomenology, Quantum Field Theory and Mathematical Physics. He has published over 100 research papers. He has received Bhatnagar award (1973), Saha award (1979), Raman medal (1996), and Goyal award (1995).He is a Fellow of all the three national academies .He is also a Fellow of TWAS, the Academy of sciences for the developing world, Trieste (1998). He was Chairman of Board of Research in Nuclear Sciences.

SANDIP P. TRIVEDI is Professor in the Department of Theoretical Physics, Tata Institute of Fundamental Research, Mumbai. His fields of specialization are String Theory and Particle Physics. He obtained his M.S. from IIT Kanpur and his Ph.D. from the California Institute of Technology, Pasadena, USA. He is a recipient of the Swarnajayanti Fellowship and the Shanti Swarup Bhatnagar Award and is a fellow of the Indian Academy of Science.